CHANGES
IN THE
LAND
Indians, Colonists, and the
Ecology of New England

CHANGES
IN THE
LAND

Indians, Colonists, and the
Ecology of New England

WILLIAM CRONON

 HILL AND WANG

A division of Farrar, Straus & Giroux

NEW YORK

Library of Congress Cataloging in Publication Data
Cronon, William.
Changes in the land.
Bibliography: p.
Includes index.
1. Man—Influence on nature—New England—History.
2. New England—History—Colonial period, ca. 1600–
1775. I. Title.
GF504.N45C76   1983   974′.02   83–7899
ISBN 0–8090–3405–0
ISBN 0–8090–0158–6 (pbk.)

For Nan

PREFACE

I have tried in this book to write an ecological history of colonial New England. By this I mean a history which extends its boundaries beyond human institutions—economies, class and gender systems, political organizations, cultural rituals—to the natural ecosystems which provide the context for those institutions. Different peoples choose different ways of interacting with their surrounding environments, and their choices ramify through not only the human community but the larger ecosystem as well. Writing a history of such relationships inevitably brings to center stage a cast of nonhuman characters which usually occupy the margins of historical analysis if they are present in it at all. Much of this book is devoted to evaluating the changing circumstances of such things as pine trees, pigs, beavers, soils, fields of corn, forest watersheds, and other elements of the New England landscape. My thesis is simple: the shift from Indian to European dominance in New England entailed important changes—well known to historians—in the ways these peoples organized their lives, but it also involved fundamental reorganizations—less well known to historians—in the region's plant and animal communities. To the cultural consequences of the European invasion— what historians sometimes call "the frontier process"—we must add the ecological ones as well. All were connected by complex relationships which require the tools of an ecologist as well as those of a historian to be properly understood.

The great strength of ecological analysis in writing history is its ability to uncover processes and long-term changes which might otherwise remain invisible. It is especially helpful in evaluating, as I do here, historical changes in modes of production: in one sense, economy in such an approach becomes a subset of ecology. I have accordingly structured my argument to take best advantage of this analytical strength. I open by contrasting the precolonial ecosystems of New England with those that existed at the beginning of the nineteenth century. I then compare the ecological relationships of precolonial Indian communities with those of the arriving Europeans, especially in terms of how the respective groups conceived of owning property (and so bounding ecosystems). Having framed the argument with these sets of contrasts, I spend the rest of the book describing the processes of ecological change that followed the Europeans' arrival.

My purpose throughout is to explain why New England habitats changed as they did during the colonial period. It is *not* my intention to rewrite the human history of the region: this is not a history of New England Indians, or of Indian-colonial relations, or of the transformation of English colonists from Puritans to Yankees. Indeed, the reader should be careful not to draw the wrong conclusions about these subjects on the basis of my text. Although I attribute much of the changing ecology of New England to the colonists' more exclusive sense of property and their involvement in a capitalist economy—both present to some extent from the 1620s onward—I do not mean to suggest that the nature of the colonial economy underwent no fundamental alterations between 1620 and 1800. It of course did, and some of those alterations, by accentuating tendencies already present, accelerated the processes of ecological change.

Equally importantly, the reader must be very clear that the Indians were no more static than the colonists in their activities and organization. When I describe precolonial Indian ways of life, I intend no suggestion that these were somehow "purer" or more "Indian" than the ways of life Indians chose (or were forced into) following their contact with colonists. Indians did not define their "Indianness" solely in terms of ecological relationships, and many of them retained their sense of identity and their resistance to colonialism even after their effective military power and political autonomy had been destroyed. Because I seek primarily to explain ecological change, I devote relatively little attention to the political and military ways in which Europeans subjugated Indian peoples. These are by now, I hope, fairly well known in their broad outline, and I trust that the reader will pursue further reading about them in the books I discuss in the bibliographical essay.

Although I have based my argument wherever possible on primary sources, any book of this kind must inevitably rely on the work of other scholars and other disciplines. Marshall Sahlins once described interdisciplinary research as "the process by which the unknowns of one's own subject are multiplied by the uncertainties of some other science." Like Sahlins, I think the benefits of interdisciplinary work outweigh the dangers, but I share his sense of risk. I have sometimes felt perilously unsure of myself as I have made my way through alien territory in anthropology, ecology, and colonial history. Fortunately, I have been blessed with guides who have pointed me clear of obvious errors whenever they could. Chief among these is Edmund S. Morgan, who originally suggested the subject of this book as a seminar paper I wrote for him four years ago. He was responsible for convincing me that the project was feasible in the first place, and has provided both criticism and moral support throughout its gestation. For a number of years now, Howard Lamar has been my mentor in all things pertaining to Western history, and I am grateful to him not only for his advice about this book but for his tolerance of my unplanned excursion onto the New England frontier.

This book could not have been written without the resources and community of Yale University. Aside from a brief excursion to Harvard's Widener Library and University Archives, all of my research was done at Yale. One of the delights (and sometimes irritations) of interdisciplinary work is the way it takes one to library call letters, library stack floors, and in fact entire libraries one has never visited before. In addition to my accustomed haunts in the Sterling and Beinecke Libraries, I found myself visiting Yale's Anthropology, Art and Architecture, Divinity School, Forestry, Geology, Kline Science, Law, Ornithology, Seeley Mudd, and Social Science Libraries; I am indebted to their librarians and to the institution which has assembled their collections. Particularly helpful was Joseph Miller, the Yale Forestry School Librarian, who took a strong interest in this project from the start.

Many other friends have helped out in a variety of ways. George Miles has been a firm and constructive critic of the way

I analyze the New England Indians, and although I am sure he continues to disagree with some of my interpretations, I have benefited a great deal from his suggestions. Timothy Weiskel has tried to keep me honest in my anthropological interpretations, and has been my chief guide to the literature of economic and ecological anthropology. Rebecca Bormann has been my most reliable and helpful ecologist critic. Others who have discussed the book with me and given me the benefit of their criticism include Jean-Christophe Agnew, Elizabeth Blackmar, John Blum, Lori Ginzberg, Fran Hallihan, Tom Hatley, David Jaffee, Tim Mitchell, Michael Saperstein, Robert Shell, Paula Shields, Barbara Smith, Gaddis Smith, Michael Smith, Robert Westbrook, and Robin Winks. To all, I give thanks.

Arthur Wang is perhaps the most congenial and encouraging publisher into whose hands a young historian could hope to fall, and I can only add my praise to the chorus one hears from his other authors. He and Eric Foner were model editors.

A special note of thanks goes to my friend David Scobey, who not only gave me an exhaustive critical reading of the manuscript in all its stages of completion but engaged in long hours of discussion about it in the midst of his own busy schedule. This book in many ways is a direct result of our extended conversations, and would have been much the worse without them.

Finally, my wife, Nan (not to mention our golden retriever, Kira), has been my companion on hikes and drives from New Haven to Cape Cod to Mount Washington, during which we have together learned most of what we know of the New England landscape. For these and other journeyings, I dedicate this book to her.

William Cronon
New Haven, Connecticut
January 1, 1983

CONTENTS

The first premise of all human history is, of course, the existence of living human individuals. Thus the first fact to be established is the physical organization of these individuals and their consequent relation to the rest of nature. . . . The writing of history must always set out from these natural bases and their modification in the course of history through the action of men.

—Karl Marx and Friedrich Engels, *The German Ideology*

As we have seen, man has reacted upon organized and inorganic nature, and thereby modified, if not determined, the material structure of his earthly home.

—George Perkins Marsh, *Man and Nature*

I think, considering our age, the great toils we have undergone, the roughness of some parts of this country, and our original poverty, that we have done the most in the least time of any people on earth.

—J. Hector St. John de Crèvecœur,
Sketches of Eighteenth-Century America

PART I

Looking Backward

1

THE VIEW
FROM WALDEN

On the morning of January 24, 1855, Henry David Thoreau sat down with his journal to consider the ways in which his Concord home had been altered by more than two centuries of European settlement. He had recently read the book *New England's Prospect*, in which the English traveler William Wood recounted his 1633 voyage to southern New England and described for English readers the landscape he had found there. Now Thoreau sought to annotate the ways in which Wood's Massachusetts was different from his own. The changes seemed sweeping indeed.[1]

He began with the wild meadow grasses, which appeared, he wrote, "to have grown more rankly in those days." If Wood's descriptions were accurate, the strawberries too had been larger and more abundant "before they were so cornered up by cultivation." Some of them had been as much as two inches around, and were so numerous that one could gather half a bushel in a forenoon. Equally abundant were gooseberries, raspberries, and especially currants, which, Thoreau mused, "so many old writers speak of, but so few moderns find wild."

New England forests had been much more extensive and their

trees larger in 1633. On the coast, where Indian settlement had been greatest, the woods had presented a more open and parklike appearance to the first English settlers, without the underbrush and coppice growth so common in nineteenth-century Concord. To see such a forest nowadays, Thoreau wrote, it was necessary to make an expedition to "the sample still left in Maine." As nearly as he could tell, oaks, firs, plums, and tulip trees were all less numerous than they had been in Wood's day.

But if the forest was much reduced from its former state, most of its tree species nevertheless remained. This was more than could be said for many of its animal inhabitants. Thoreau's list of those that were now absent was stark: "bear, moose, deer, porcupines, 'the grim-fac'd Ounce, and rav'nous howling Wolf,' and beaver. Martens." Not only the mammals of the land were gone; the sea and air also seemed more empty. Bass had once been caught two or three thousand at a time. The progeny of the alewives had been "almost incredible." Neither was now present in such abundance. Of the birds, Thoreau wrote: "Eagles are probably less common; pigeons of course . . . heath cocks all gone . . . and turkeys . . . Probably more owls then, and cormorants, etc., etc., sea-fowl generally . . . and swans." To Wood's statement that one could purchase a fresh-killed swan for dinner at the price of six shillings, Thoreau could only write in wonderment, "Think of that!"

There is a certain plaintiveness in this catalog of Thoreau's, a romantic's lament for the pristine world of an earlier and now lost time. The myth of a fallen humanity in a fallen world is never far beneath the surface in Thoreau's writing, and nowhere is this more visible than in his descriptions of past landscapes. A year after his encounter with William Wood's New England of 1633, he returned to its lessons in more explicitly moral language. "When I consider," he wrote, "that the nobler animals have been exterminated here,—the cougar, panther, lynx, wolverene, wolf, bear, moose, deer, the beaver, the turkey, etc., etc.,—I cannot but feel as if I lived in a tamed, and, as it were, emasculated country." Seen in this way, a changed landscape meant a loss of wildness and virility that was ultimately spiritual in its import, a sign of declension in both nature and humanity. "Is it not," Thoreau asked, "a maimed and imperfect nature that I am conversant with?"[2]

It is important that we answer this question of Thoreau's carefully: how did the "nature" of New England change with the coming of the Europeans, and can we reasonably speak of its changes in terms of maiming and imperfection? There is nothing new to the observation that European settlement transformed the American landscape. Long before Thoreau, naturalists and historians alike were commenting on the process which was converting a "wilderness" into a land of European agricultural settlement. Whether they wrote of Indians, the fur trade, the forest, or the farm, colonial authors were constantly aware that fundamental alterations of the ecological fabric were taking place around them. The awareness increased as time went on. By the late eighteenth century, many individuals—Peter Kalm, Peter Whitney, Jeremy Belknap, and Timothy Dwight chief among them—were commenting extensively on these changes.

For the most part, unlike Thoreau, they did so approvingly. As early as 1653, the historian Edward Johnson could count it as one of God's providences that a "remote, rocky, barren, bushy, wild-woody wilderness" had been transformed in a generation into "a second England for fertilness." In this vision, the transformation of wilderness betokened the planting of a garden, not the fall from one; any change in the New England environment was divinely ordained and wholly positive. By the end of the eighteenth century, the metaphors for environmental change had become more humanistic than providential, but were no less enthusiastic about the progress such change represented. In a passage partially anticipating Frederick Jackson Turner's frontier thesis, for instance, Benjamin Rush described a regular sequence for clearing the forest and civilizing the wilderness. "From a review [of] the three different species of settlers," he wrote, speaking of Pennsylvania, "it appears, that there are certain regular stages which mark the progress from the savage to civilized life. The first settler is nearly related to an Indian in his manners— In the second, the Indian manners are more diluted: It is in the third species of settlers only, that we behold civilization completed." Though landscape was altered by this supposed social evolution, the *human* process of development—from Indian to clearer of the forest to prosperous farmer—was the center of Rush's attention. Environmental change was of secondary interest. For Enlightenment thinkers like Rush, in each stage, the shape of the landscape was a visible confirmation of the state of human society. Both underwent an evolutionary development from savagery to civilization.[3]

Whether interpreted as declension or progress, the shift from Thoreau's forest of "nobler animals" to Rush's fields and pastures of prosperous farmers signaled a genuinely transformed countryside, one whose changes were intimately bound to the human history which had taken place in its midst. The replacement of Indians by predominantly European populations in New England was as much an ecological as a cultural revolution, and the human side of that revolution cannot be fully understood until it is embedded in the ecological one. Doing so requires a history, not only of human actors, conflicts, and economies, but of ecosystems as well.

How might we construct such an ecological history? The types of evidence which can be used to evaluate ecological change before 1800 are not uniformly reliable, and some are of a sort not ordinarily used by historians. It is therefore important to reflect on how they should best be criticized and used. The descriptions of travelers and early naturalists, for instance, provide observations of what New England looked like in the early days of European settlement, and how it had changed by the end of the eighteenth century. As such, they provide the backbone of this study. But to use them properly requires that we evaluate each traveler's skills as a naturalist, something for which there is often only the evidence of his or her writings. Moreover, we can only guess at how ideological commitments such as Thoreau's or Rush's colored the ways they saw the landscape. How much did William Wood's evident wish to promote the Massachusetts Bay Colony lead him to idealize its environment? To what extent did the anonymous author of *American Husbandry* shape his critique of American agriculture to serve his purpose of preserving colonial attachments to Britain? Even if we can remove most of these ideological biases to discover what it was a traveler actually saw, we must still acknowledge that each traveler visited only a tiny fraction of the region. As Timothy Dwight once remarked, "Your travelers seize on a single person, or a solitary fact, and make them the representatives of a whole community and a general custom." We are always faced with the problem of generalizing from a *local* description to a *regional* landscape, but our understanding of modern ecosystems can be of great help in doing so.[4]

A second fund of data resides in various colonial town, court, and legislative records, although here the evidence of ecological change can sometimes be tantalizingly elliptical. We cannot always know with certainty whether a governmental action anticipated or reacted to a change in the environment. When a law was passed protecting trees on a town commons, for example, did this mean that a timber shortage existed? Or was the town merely responding with prudent foresight to the experience of other localities? If a shortage existed, how severe was it? Was it limited only to certain species of trees? And so on. Only by looking at the overall pattern of legal activity can we render a reasonable judgment on such questions. These problems notwithstanding, town and colony records address almost the entire range of ecological changes in colonial New England: deforestation, the keeping of livestock, conflicts between Indians and colonists over property boundaries, the extermination of predators such as wolves, and similar matters. Deeds and surveyor records can be used statistically to estimate the composition of early forests, and are usually more accurate than travelers' accounts even though subject to sampling errors.[5]

Then there are the less orthodox sorts of evidence which historians borrow from other disciplines and have less experience in criticizing. Relict stands of old-growth timber, such as the Cathedral Pines near Cornwall, Connecticut, can suggest what earlier forests may have looked like. The relict stands which exist today, however, are by no means identical to most of the forests which existed in colonial times, so that the record of earlier forests must be sought in less visible places. Ecologists have done very creative detective work in analyzing tree rings, charcoal deposits, rotting trunks, and overturned stumps to determine the history of several New England woodlands. The fossil pollen in pond and bog sediments is a reliable but fuzzy indicator of the changing species composition of surrounding vegetation; despite problems in determining the absolute age of such pollen, it supplies some of the most reliable evidence for reconstructing past forests. In addition, a wide variety of archaeological evidence can be used to assess past environments, particularly the changing relations of human inhabitants to them.[6]

Finally, there are those awkward situations in which an ecological change which undoubtedly must have been occurring in the colonial period has left little or no historical evidence at all. These include microscopic changes in soil fauna and flora, soil compaction, changes in the transpiration rates of forests, and so on. Here all arguments become somewhat speculative. Given what we know of ecosystem dynamics, however, it would be wrong simply to ignore such changes, since some of them almost certainly occurred even though no one noticed them at the time. I will occasionally appeal to modern ecological literature to assert the probability of such changes, and trust that, by so doing, I am not straying too far from the historian's usual canon for evaluating evidence. Silences in the historical record sometimes require us to make the best-informed interpolations we can, and I have tried always to be conservative on the few occasions when I have been forced to do this.

Although caution is required in handling all these various forms of evidence (and nonevidence), together they provide a remarkably full portrait of ecological change in colonial New England. But they also raise intriguing questions, questions which are both empirical and theoretical. One, for instance, follows directly from the imprecision of the data: travelers' accounts and other colonial writings are not only subjective but often highly generalized. Colonial nomenclature could be quite imprecise, so much so that the French traveler Chastellux wrote of the impoverishment of American English as a result:

> Anything that had no English name has here been given only a simple designation: the jay is the blue bird, the cardinal the red bird; every water bird is simply a duck, from the teal to the wood duck, and to the large black duck which we do not have in Europe. They call them "red ducks," "black ducks," "wood ducks." It is the same with respect to their trees: the pine, the cypresses, the firs, are all included under the general name of "pine trees."

More confusing still could be the natural tendency for colonists to apply European names to American species which only superficially resembled their counterparts across the ocean.[7]

The problems which this fuzzy nomenclature can create for

those doing ecological history should be obvious. For instance, many early descriptions, including those by William Wood, make no mention at all of hemlock, although they do mention fir and spruce. On just such evidence, Thoreau concluded—incorrectly—that the fir tree had been much more common in colonial times. But since fir and spruce are now largely absent in southern New England, and since fossil pollen shows that hemlock has long been a significant component of the New England forest, it seems reasonable to conclude that "hemlock" was subsumed by colonists under the names of "fir," "spruce," and probably "pine." But how common was it? Only the fossil pollen can tell us. As another example, the hickory was rarely mentioned by name, but instead was for a long time known as the "walnut," an entirely different genus of tree. Because white pine was valuable economically, many early visitors seem to have seen it everywhere, thus leading them to exaggerate its numerical significance. Colonists confused the native junipers with European cedars for the same economic reasons, so that the red cedar has carried a misleading name ever since. All of these problems of nomenclature must be borne in mind if one is not to give a distorted picture of the colonial ecosystem.[8]

A second difficulty is the old and familiar fallacy of *post hoc ergo propter hoc.*[9] When reading colonial accounts describing floods, insect invasions, coastal alterations, and significant changes in climate, we are perhaps all too tempted to attribute these by some devious means to the influence of the arriving Europeans. This will not always do. Not all the environmental changes which took place after European settlement were caused by it. Some were part of much longer trends, and some were random: neither type need have had anything to do with the Europeans. Trickier still are instances where Europeans may or may not have altered the *rate* at which a change was already occurring. Unless one can show some plausible mechanism whereby settlement could and probably did cause a change, it seems best not to attribute it to European influence. One cannot escape the fallacy altogether— any discussion of causality in history must encounter it with some frequency—but one must at least be aware of when one is flirting with it. I shall have occasion to do so here.

This brings us to the heart of the theoretical difficulties involved in doing ecological history. When one asks how much an ecosystem has been changed by human influence, the inevitable next question must be: "changed in relation to what?" There is no simple answer to this. Before we can analyze the ways people alter their environments, we must first consider how those environments change in the absence of human activity, and that in turn requires us to reflect on what we mean by an ecological "community." Ecology as a biological science has had to deal with this problem from its outset. The first generation of academic ecologists, led by Frederic Clements, defined the communities they studied literally as superorganisms which experienced birth, growth, maturity, and sometimes death much as individual plants and animals did. Under this model, the central dynamic of community change could be expressed in the concept of "succession." Depending on its region, a biotic community might begin as a pond, which was then gradually transformed by its own internal dynamics into a marsh, a meadow, a forest of pioneer trees, and finally to a forest of dominant trees. This last stage was assumed to be stable and was known as the "climax," a more or less permanent community which would reproduce itself indefinitely if left undisturbed. Its equilibrium state defined the mature forest "organism," so that all members of the community could be interpreted as functioning to maintain the stability of the whole. Here was an apparently objective point of reference: any actual community could be compared with the theoretical climax, and differences between them could then usually be attributed to "disturbance." Often the source of disturbance was human, implying that humanity was somehow outside of the ideal climax community.[10]

This functionalist emphasis on equilibrium and climax had important consequences, for it tended to remove ecological communities from history. If all ecological change was either self-equilibrating (moving toward climax) or nonexistent (remaining in the static condition of climax), then history was more or less absent except in the very long time frame of climatic change or Darwinian evolution. The result was a paradox. Ecologists trying to define climax and succession for a region like New England were faced with an environment massively altered by human beings, yet their research program demanded that they determine what that environment would have been like without a human presence. By peeling away the corrupting influences of man and woman, they could discover the original ideal community of the climax. One detects here a certain resemblance to Thoreau's reading of William Wood: historical change was defined as an aberration rather than the norm.[11]

In time, the analogy comparing biotic communities to organisms came to be criticized for being both too monolithic and too teleological. The model forced one to assume that any given community was gradually working either to become or to remain a climax, with the result that the dynamics of nonclimax communities were too easily ignored. For this reason, ecology by the mid-twentieth century had abandoned the organism metaphor in favor of a less teleological "ecosystem." Now individual species could simply be described in terms of their associations with other species along a continuous range of environments; there was no longer any need to resort to functional analysis in describing such associations. Actual relationships rather than mystical superorganisms could become the focus of study, although an infusion of theory from cybernetics encouraged ecologists to continue their interest in the self-regulating, equilibrating characteristics of plant and animal populations.[12]

With the imperatives of the climax concept no longer so strong, ecology was prepared to become at least in part a historical science, for which change was less the result of "disturbance" than of the ordinary processes whereby communities maintained and transformed themselves. Ecologists began to express a stronger interest in the effects of human beings on their environment. What investigators had earlier seen as an inconvenient block to the discovery of ideal climax communities could become an object of research in its own right. But accepting the effects of human beings was only part of this shift toward a more historical ecology. Just as ecosystems have been changed by the historical activities of human beings, so too have they had their own less-recorded history: forests have been transformed by disease, drought, and fire, species have become extinct, and landscapes have been drastically altered by climatic change without any human intervention at all. As we shall see, the period of human occupation in postglacial New England has seen environmental changes on an enormous scale, many of them wholly apart from human influence. There has been no timeless wilderness in a state of perfect changelessness, no climax forest in permanent stasis.

But admitting that ecosystems have histories of their own still leaves us with the problem of how to view the people who inhabit them. Are human beings inside or outside their systems? In trying to answer this question, appeal is too often made to the myth of a golden age, as Thoreau sometimes seemed inclined to do. If the nature of Concord in the 1850s—a nature which many Americans now romanticize as the idyllic world of Thoreau's own Walden—was as "maimed" and "imperfect" as he said, what are we to make of the wholeness and perfection which he thought preceded it? It is tempting to believe that when the Europeans arrived in the New World they confronted Virgin Land, the Forest Primeval, a wilderness which had existed for eons uninfluenced by human hands. Nothing could be further from the truth. In Francis Jennings's telling phrase, the land was less virgin than it was widowed. Indians had lived on the continent for thousands of years, and had to a significant extent modified its environment to their purposes. The destruction of Indian communities in fact brought some of the most important ecological changes which followed the Europeans' arrival in America. The choice is not between two landscapes, one with and one without a human influence; it is between two human ways of living, two ways of belonging to an ecosystem.[13]

The riddle of this book is to explore why these different ways of living had such different effects on New England ecosystems. A group of ecological anthropologists has tried to argue that for many non-Western societies, like those of the New England Indians, various ritual practices have served to stabilize people's relationships with their ecosystems. In effect, culture in this anthropological model becomes a homeostatic, self-regulating system much like the larger ecosystem itself. Thus have come the now famous analyses designed to show that the slaughter of pigs in New Guinea, the keeping of sacred cows in India, and any number of other ritual activities, all function to keep human populations in balance with their ecosystems. Such a view would describe precolonial New England not as a virgin landscape of natural harmony but as a landscape whose essential characteristics were kept in equilibrium by the cultural practices of its human community.[14]

Unfortunately, this functional approach to culture has the same penchant for teleology as does the organism model of eco-

logical climax. Saying that a community's rituals and social institutions "function" unconsciously to stabilize its ecological relationships can lead all too quickly into a static and ahistorical view
of both cultural agency and ecological change. If we assume *a
priori* that cultures are systems which tend toward ecological
stability, we may overlook the evidence from many cultures—
even preindustrial ones—that human groups often have significantly *unstable* interactions with their environments. When we
say, for instance, that the New England Indians burned forests
to clear land for agriculture and to improve hunting, we describe
only what they themselves thought the purpose of burning to be.
But to go further than this and assert its unconscious "function"
in stabilizing Indian relationships with the ecosystem is to deny
the evidence from places like Boston and Narragansett Bay that
the practice could sometimes go so far as to remove the forest
altogether, with deleterious effects for trees and Indians
alike.[15]

All human groups consciously change their environments to
some extent—one might even argue that this, in combination
with language, is the crucial trait distinguishing people from
other animals—and the best measure of a culture's ecological
stability may well be how successfully its environmental changes
maintain its ability to reproduce itself. But if we avoid assumptions about environmental equilibrium, the *instability* of human
relations with the environment can be used to explain both cultural and ecological transformations. An ecological history begins by assuming a dynamic and changing relationship between
environment and culture, one as apt to produce contradictions as
continuities. Moreover, it assumes that the interactions of the
two are dialectical. Environment may initially shape the range of
choices available to a people at a given moment, but then culture
reshapes environment in responding to those choices. The re-
shaped environment presents a new set of possibilities for cultural reproduction, thus setting up a new cycle of mutual deter-
mination. Changes in the way people create and re-create their
livelihood must be analyzed in terms of changes not only in their
social relations but in their *ecological* ones as well.

Doing away with functionalism does not mean abandoning the
system-oriented perspective which an ecological approach allows. In addition to assuming that relations between people and

their environment are not constant, but rather historical and dialectical, it sees those relations as being connected within an interacting system. Efforts to describe ecological history simply in terms of the transfer of individual species between segregated ecosystems, as Alfred Crosby and William H. McNeill have done, are thus bound to be incomplete.[16] Important as organisms like smallpox, the horse, and the pig were in their direct impact on American ecosystems, their full effect becomes visible only when they are treated as integral elements in a complex system of environmental and cultural relationships. The pig was not merely a pig but a creature bound among other things to the fence, the dandelion, and a very special definition of property. It is these kinds of relationships, the contradictions arising from them, and their changes in time, that will constitute an ecological approach to history.

The study of such relations is usually best done at the local level, where they become most visible; the best ecological histories to date have all examined relatively small systems as cases. I have opted for a similar approach, albeit for a somewhat larger region. But despite its strengths, the choice of a small region has one crucial problem: how do we locate its boundaries? Traditionally in anthropology, this has simply involved the area within which people conduct their subsistence activities, often described using "ethno-ecological" techniques which analyze the way the inhabitants themselves conceive of their territory.[17] Yet anthropologists are increasingly aware, as historians have long known, that the development of a world capitalist system has brought more and more people into trade and market relations which lie well beyond the boundaries of their local ecosystems. Explaining environmental changes under these circumstances— as in the shift from Indian to European populations in colonial New England—becomes even more complex than explaining changes internal to a local ecosystem. In an important sense, a distant world and its inhabitants gradually become part of another people's ecosystem, so that it is increasingly difficult to know which ecosystem is interacting with which culture. This erasure of boundaries may itself be the most important issue of all.

In colonial New England, two sets of human communities which were also two sets of ecological relationships confronted each other, one Indian and one European. They rapidly came to inhabit a single world, but in the process the landscape of New England was so transformed that the Indians' earlier way of interacting with their environment became impossible. The task before us is not only to describe the ecological changes that took place in New England but to determine what it was about Indians and colonists—in their relations both to nature and to each other—that brought those changes about. Only thus can we understand why the Indian landscape of precolonial times had become the much altered place Thoreau described in the nineteenth century.

The view from Walden in reality contained far more than Thoreau saw that January morning in 1855. Its relationships stretched beyond the horizons of Concord to include vistas of towns and markets and landscapes that were not in Thoreau's field of vision. If only for this reason, we must beware of following him too closely as our guide in these matters. However we may respect his passion, we must also recognize its limits:

> I take infinite pains to know all the phenomena of the spring, for instance, thinking that I have here the entire poem, and then, to my chagrin, I hear that it is but an imperfect copy that I possess and have read, that my ancestors have torn out many of the first leaves and grandest passages, and mutilated it in many places. I should not like to think that some demigod had come before me and picked out some of the best of the stars. I wish to know an entire heaven and an entire earth.

We may or may not finally agree with Thoreau in regretting the changes which European settlers wrought in the New World, but we can never share his certainty about the possibility of knowing an entire heaven and an entire earth. Human and natural worlds are too entangled for us, and our historical landscape does not allow us to guess what the "entire poem" of which he spoke might look like. To search for that poem would in fact be a mistake. Our project must be to locate a nature which is within rather than without history, for only by so doing can we find human communities which are inside rather than outside nature.[18]

PART II

The Ecological

Transformation of

Colonial New England

2

LANDSCAPE AND
PATCHWORK

The land as the Europeans found it differed in key ways from the one they had left behind, but their understanding of *how* it differed only emerged gradually. Their image of the area that came to be called New England was shaped by a variety of circumstances that had little to do with the ecology of the region itself. The first European visitors, for instance, saw only areas within easy reach of the coast. For the entirety of the sixteenth century, maps of New England consisted of a single line separating ocean from land, accompanied by a string of place-names to indicate landmarks along the shore; the interior remained blank. Verbal descriptions were not even this thorough. At a handful of coastal points, explorers like Verrazzano, Gosnold, Pring, and Champlain made landfalls that were eventually written up in a paragraph or two, but such accounts were few and far between. Only when settlement began in the 1620s did fuller descriptions start to appear, and even then they were limited to areas within a few miles of the coast or along a few major rivers. For many years, the only New England known to Europe was near salt water.[1]

Once European visitors had arrived, their preconceptions and expectations led them to emphasize some elements of the landscape and to filter out certain others. Most of the early explorers sought to discover what Richard Hakluyt had called "merchantable commodities" in his classic *Discourse Concerning Western Planting* in 1584. These were the natural products which could be shipped to Europe and sold at a profit in order to provide a steady income for colonial settlements. Theoreticians of colonialism like Hakluyt had furnished a ready list of such commodities by the time Europeans began to visit New England regularly: fish for salting, furs for clothing, timber for ships, sassafras for curing syphilis, and so on. Visitors inevitably observed and recorded greater numbers of "commodities" than other things which had not been labeled in this way. It was no accident that James Rosier referred to the coastal vegetation of Maine as "the profits and fruits which are naturally on these Ilands." His word "profits" may not have connoted the marginal gain from a mercantile transaction, but it did identify those natural products which were of potential use to a European way of life. Descriptions framed on such a basis were bound to say as much about the markets of Europe as they did about the ecology of New England.[2]

Distorting as this emphasis on commodities might be to an accurate understanding of the New England environment, it helps explain one reason why Europeans found the American landscape so different from the one they had left behind. What was a "merchantable commodity" in America was what was scarce in Europe. Only if this was true would it make sense to pay the cost of transporting it across the ocean. Beaver, cod, and sassafras all satisfied this economic requirement and so were often the chief goals of an exploring expedition. But even something like firewood, which was too bulky to justify trans-Atlantic shipment and could thus only be used by settlers in America itself, might be perceived as a commodity because of its scarcity value in England. England had been experiencing a near crisis in its wood supply since at least the time of Columbus, with the result that this single most important source of heating and building materials became increasingly costly throughout the century preceding the English revolution. Parliament began to restrict the cutting of English timber as early as 1543. By the time

settlement began in New England, coal production had started
to rise in the fields of Durham and Northumberland, and London
was beginning its dependence on the fuel that would soon make
it renowned for its terrible fogs. Even if explorers and settlers
could not initially ship American timber back home, their awareness of the English wood scarcity colored the way they reacted
to New England forests. As often as not, their descriptions of
New England contained implicit comparisons with England.[3]

Seeing landscapes in terms of commodities meant something
else as well: it treated members of an ecosystem as isolated and
extractable units. Explorers describing a new countryside with
an eye to its mercantile possibilities all too easily fell into this
way of looking at things, so that their descriptions often degene-
rated into little more than lists. Martin Pring's account of the
trees of Martha's Vineyard illustrates this tendency:

> As for Trees the Country yeeldeth Sassafras a plant of sove-
> reigne vertue for the French Poxe, and as some of late have
> learnedly written good against the Plague and many other
> Maladies; Vines, Cedars, Okes, Ashes, Beeches, Birch trees,
> Cherie trees bearing fruit whereof wee did eate, Hasels,
> Wichhasels, the best wood of all other to make Sope-ashes
> withall, walnut-trees, Maples, holy to make Bird lime with,
> and a kinde of tree bearing a fruit like a small red Peare-
> plum.

Little sense of ecological relationships emerges from such a list.
One could not use it to describe what the forest actually looked
like or how these trees interacted with one another. Instead, its
purpose was to detail resources for the interest of future un-
dertakings.[4]

How these resources were perceived depended a great deal on
whether one contemplated an expedition that would simply
gather the "profits" of a countryside or one that would settle a
new plantation. Settlers who had actually to live in a New World
environment were less likely than their merchant companions to
view it as a linear list of commodities. Their very survival required that they manipulate the environment, and so it is from
their writings that a sense of ecological relationships begins to
emerge. Settlers had first to survive and prosper before they

could sell commodities across the sea, and that meant understanding the land they lived in. By the time they did this, however, the land was already changing in response to that new
understanding, creating a landscape different from the one that
had been there before.[5]

All these things—the limited areas visited by Europeans, their
tendency to view the landscape in terms of their own cultural
concepts, their selective emphasis on commodities, the ecological
changes they themselves wrought—meant that their record of
precolonial New England ecosystems was inevitably incomplete.
The fragments they have left us testify as much to their own
cultural preconceptions as to the actual environments they encountered. But there was one European perception that was undoubtedly accurate, and about it all visitors were agreed—the
incredible abundance of New England plant and animal life, an
abundance which, when compared with Europe, left more than
one visitor dumbfounded. Many found themselves protesting to
correspondents on the other side of the Atlantic that, however
hard it was to believe, they were not exaggerating their reports
of what they had discovered there.

The experience of New England's plenty began with the fish
of the coastal waters, which had been the original reason that
Breton, Portuguese, and Bristol fishermen had started visiting
the area in the fifteenth century. "The aboundance of Sea-Fish,"
wrote the Reverend Francis Higginson in 1630, "are almost beyond beleeving, and sure I should scarce have beeleved it except
I had seene it with mine owne eyes." John Brereton described
how, in a few hours of fishing, he and his companions "had
pestered our ship so with Cod fish, that we threw numbers of
them over-boord againe." Cape Cod came to be named as a result
of such experiences, which nearly all of the early explorers mention. But the real statements of wonder came from visitors to the
settlements, who saw the spring spawning runs of smelt, alewives, sturgeon, and other ocean fish which migrated to fresh
water to deposit their eggs. William Wood described the arrival
of the alewives "in such multitudes as is almost incredible, press-
ing up such shallow waters as will scarce permit them to swim."
So thick did the fish become in some streams that at least one
inhabitant fancied he might have walked on their backs without
getting his feet wet. John Josselyn had no illusions about crossing

streams on the backs of fish, but he was sure that he could have walked knee-deep through stranded herring across a quarter mile of beach. Nothing in their English experience prepared these men for the sight of such prodigious quantities of fish.[6]

The same was true of the region's birds. Wood hesitated to describe how easy it was to hunt waterfowl in New England. "If I should tell you," he wrote, "how some have killed a hundred geese in a week, fifty ducks at a shot, forty teals at another, it may be counted impossible though nothing more certain." Such birds were present in greatest numbers during the spring and fall migrations, but others, like the turkey, could be hunted year-round. Not only did the wild turkeys seem fatter and sweeter than the domesticated turkeys of Europe—which few colonists even remembered had once been imported from the New World —but their behavior could hardly have been better suited to those who sought to hunt them. As Thomas Morton described it, they were easily shot "because, the one being killed, the other sit fast nevertheless; and this is no bad commodity." A man could kill a dozen turkeys in half a day.[7]

For sheer abundance, though, only one bird could match the alewives. Nothing so astonished Europeans about New England as the semiannual flights of the passenger pigeons. John Josselyn measured their numbers in the "millions of millions," and spoke of flocks "that to my thinking had neither beginning nor ending, length nor breadth, and so thick that I could see no Sun." Thomas Dudley told of a March day in 1631 when "there flew over all the towns in our plantations . . . many flocks of doves, each flock containing many thousands and some so many that they obscured the light." Again settlers felt the need to protest their honesty as they wrote descriptions of this kind. "Those that did not see them," said one, "might think it was not true, but it is very true."[8]

None of the mammals reproduced themselves in such concentrated numbers, but they too impressed English visitors accustomed to a landscape in which much of the available hunting was reserved to large landowners and the Crown. "For Beasts," wrote Higginson, "there are some Beares . . . Also here are severall sorts of Deere . . . Also Wolves, Foxes, Beavers, Otters, Martins, great wild Cats, and a great Beast called a Molke [moose] as bigge as an Oxe." Thomas Morton found New England's deer—among which he included elk—to be larger than English fallow deer, and regarded them as "the most usefull and most beneficiall beast" of the region. In spring, one could see as many as a hundred of them in the space of a mile, and they were numerous enough at other times to supply meat year-round. Still, one of the earliest ecological relationships of which the colonists were aware led them to believe that the numbers of deer might be increased if only wolves could be eliminated. "Here is good store of deer," wrote William Hammond; "were it not for the wolves here would be abound, for the does have most two fawns at once, and some have three, but the wolves destroy them."[9]

Visitors and colonists were as impressed by the animals that were absent from New England as by those that were present. Some were familiar wild species, many closely associated with human settlements, whose ranges did not reach to the New World: magpies, cuckoos, nightingales, larks, and sparrows were all missed by the colonists. More striking was the absence of the domesticated animals—horses, sheep, goats, swine, cats, and cattle—which arrived only after 1620. The only indigenous dogs were near kin of the wolf, and although mice were common, there were no rats. "But for Rats," wrote Morton, "the Country by Nature is troubled with none." A number of microscopic organisms were absent as well, but these were commented on mainly in terms of the colonists' remarkable healthiness. Several inhabitants agreed with the observation: "For the common diseases of England, they be strangers to the English now in that strange land. To my knowledge I never knew any that had the pox, measles, green-sickness, headaches, stone, or consumptions, etc." Disease was by no means absent from New England, as deaths from "seasoning" and epidemics both showed, but the colonial population nevertheless remained for a while relatively isolated from the European disease environment. Large numbers of deaths in the occasional epidemics which did occur should not obscure the fact that New England mortality rates—for Europeans—were on average much lower than comparable rates in Europe.[10]

New England's abundance was not confined to its animal inhabitants. Indeed, English settlers accustomed to scarcities of wood were perhaps most delighted by the forests they found there. Here, wrote one visitor to Plymouth, was "good ground in abundance, with excellent good timber." William Wood, whose description of the woods around Boston would later so intrigue Thoreau, furnished a more precise picture of the Massachusetts forest. "The timber of the country," he wrote, "grows straight and tall, some trees being twenty, some thirty foot high, before they spread forth their branches." The most common species in southern New England were oaks, hickories, chestnuts, and pines. If anyone doubted what such trees meant to an Englishman, Francis Higginson made the matter clear: they meant being warm in winter, warmer even than the nobility of England could hope to be.

> Though it bee here somewhat cold in the winter, yet here we have plenty of Fire to warme us, and that a great deale cheaper then they sel Billets and Faggots in *London:* nay, all *Europe* is not able to afford so great Fires as *New-England.* A poor servant here that is to possesse but 50 Acres of land, may afford to give more wood for Timber and Fire as good as the world yeelds, then many Noble men in *England* can afford to do.

Higginson's conclusion said as much about the English fuel crisis as it did about New England's forests: "Here is good living for those that love good Fires."[11]

One must not visualize the New England forest at the time of settlement as a dense tangle of huge trees and nearly impenetrable underbrush covering the entire landscape. Along the southern coast, from the Saco River in Maine all the way to the Hudson, the woods were remarkably open, almost parklike at times. When Verrazzano visited Narragansett Bay in 1524, he found extensive open areas and forests that could be traversed easily "even by a large army." A century later, William Wood made similar observations about Massachusetts Bay. "Whereas it is generally conceived that the woods grow so thick that there is no more clear ground than is hewed out by labor of man," he wrote, "it is nothing so, in many places diverse acres being clear so that one may ride ahunting in most places of the land if he will venture himself for being lost." At a number of sites, trees were entirely absent. Higginson spoke of a hill near Boston from which one could see "thousands of acres" with "not a Tree in the same." Boston itself was in fact nearly barren, and colonists were forced to seek wood from nearby islands.[12]

In coastal areas north of the Saco, and in the mountainous interior of present-day New Hampshire and Vermont, the forest became less open and its composition changed. When Verrazzano made landfall in Maine, he found "high country full of very dense forests, composed of pines, cypresses, and similar trees which grow in cold regions." Later visitors concurred that the forests of northern New England were denser, often more coniferous, and, for all of their magnificence, generally less hospitable than those of the south. Thomas Morton described spruce trees in these cold northern woods that measured as much as twenty feet around, and Josselyn mentioned some that were so big that no ship could carry them. The farther north one traveled in New England, the colder the climate became: snow stayed on the ground a month or two longer in the interior of Maine than it did in Massachusetts, and the frost-free growing season fell from about 200 days in southern Connecticut to just over 150 days on the coast of Maine. The failure of the Popham colony at Sagadahoc in 1608 helped create an unfavorable image for much of northern New England. By 1624, John Smith could describe the coast north of the Penobscot River as "a Countrey rather to affright then delight one, and how to describe a more plaine spectacle of desolation, or more barren, I know not." By way of contrast, Smith regarded Massachusetts Bay as "the Paradice of all those parts," suggesting that, however fragmentary their knowledge, he and other colonial observers were well aware of the diversity of New England environments.[13]

Ecologists have traditionally divided New England into several vegetational zones which reflect these broad differences between northern and southern forests. Thus, the south, including all of Connecticut, Rhode Island, and the eastern fourth of Massachusetts, was known as the "oak-chestnut" region before the chestnut was destroyed by blight in the early part of the twentieth century. In colonial times, the area was dominated by a variety of "central hardwoods"—black, red, and white oaks, chestnut, and the hickories—in addition to hemlock and scattered stands of white pine. In much of the north, on the other hand, including most of Vermont, the northern two-thirds of New Hampshire, and almost all of Maine, "northern hardwoods" such as beech, yellow birch, and the maples predominated, with red spruce and balsam fir occurring at higher elevations and in swamps. Between the north and the south—in western Massachusetts, southeastern New Hampshire, and along the Connecticut River north of Springfield—there was a zone of transition that contained significant mixtures of both northern and southern New England species, where the full range of hardwoods joined with white pine and hemlock to create a dense, moist forest.[14]

The trouble with such zones is that although they demarcate large-scale regions, they obscure as much as they reveal. The precolonial forest was a mosaic of tree stands with widely varying compositions. Each individual tree species had its own unique range and ecological characteristics, so that many different combinations of species could be found within a single vegetational zone or even within a few square miles. In 1605, James Rosier told of walking up a river in Maine—in the "northern hardwood zone"—and finding a forest which nevertheless consisted of great old oaks growing widely scattered in open fields, with occasional birches, hazels, and strawberries mixed in. From time to time, his company passed through "lowe Thicks" of dense young shrubs and saplings, made up of still other species. On the three hills they climbed, they found "high timber trees," presumably spruce or pine, which were fit to serve as "masts for ships of 400 tun." As Rosier described the place, "It did all resemble a stately Parke, wherein appeare some old trees with high withered tops, and other flourishing with living green boughs." In the space of a mere four miles, Rosier and his men had encountered several very different forests arranged in a complex patchwork upon the landscape.[15]

This kind of diversity was typical of the New England landscape, and is at least as important as larger vegetational zones to the way we should understand the ecology of that landscape. Drainage patterns, the hilliness of the ground, the range of soils, the nature of the bedrock, the location of Indian settlements—all played important roles in determining what vegetation and animal life existed where. These influences applied to more than just the precolonial forest: Francis Higginson's interest was not merely academic when he described the different soils of Massachusetts Bay. "It is a land," he wrote, "of divers and sundry sorts all about *Masathulets* Bay, and at *Charles* River is as fat blacke Earth as can be seene anywhere: and in other places you have a clay soyle, in other gravell, in other sandy, as it is all about our Plantation at *Salem.* " The nature and diversity of an area's soils might be crucial to the future prosperity of a new settlement, determining the success or failure of its agriculture. As we shall see, colonists studied the native trees carefully for indications of soil fertility.[16]

Even in a relatively small area like eastern Massachusetts, it was possible to find a remarkable range of different habitats. Although colonists generally described the forest as an open oak woodland, there were many poorly drained sites in lowland places, whether along streams or in swamps, where red maple, swamp white oak, alders, and willows were the principal vegetation. William Wood described some of these areas as being twenty or thirty miles in extent. He noted that their watercourses often preserved large areas from the fires—many of them set by Indians—that cleared the underbrush elsewhere, leaving thickets through which it was nearly impossible for a traveler to pass. The thickets offered excellent refuge for deer, and surrounding areas were often prime hunting places. The Indians referred to such lowlands as "abodes of owls," and used them as hiding places during times of war.[17]

An entirely different wetland habitat occasionally occurred where a dense mat of sphagnum moss, leatherleaf, and various sedges grew out from the edges of a pond. Usually the mat was underlaid by water, so that people jumping up and down on it could feel the earth move beneath them like a giant waterbed. Indians referred to such areas as places "where the earth shakes and trembles"; the English called them "quaking bogs." Plants that could grow in these highly acidic environments had to be adapted to a water world of few nutrients and little oxygen. Many such plants grew nowhere else: cranberries, which the colonists eventually came to appreciate, the parasitic orchids, and the insect-consuming sundews and pitcher plants. From the colonists' initial point of view, the most attractive feature of the bogs was the Atlantic white cedars that grew around their edges. As Morton wrote, "If any man be desirous to finde out in what part of the Country the best Cedars are, he must get into the bottom grounds, and in vallies that are wet at the spring of the yeare."

When a bog was finally overgrown by vegetation, a dense stand of white cedar and red maple might be the only visible sign of its passing. Many of the places Morton described were probably of this sort, a particularly ephemeral habitat that was easily subject to human influence.[18]

At the opposite end of the spectrum were soils which were so well drained that they created very dry conditions for the trees living on them. The sandy soils and glacial tills of Cape Cod were the most extensive examples of this, but smaller ones occurred on the sandy outwash plains of some rivers and in rare sand barrens like those of North Haven, Connecticut. Although Cape Cod possesses the mildest and most temperate climate in New England, with 44 inches of rain and over 210 frost-free days annually, its typical forest is made up of scrubby trees adapted to extreme dryness. Chief among these are the pitch pines, deeply rooted trees which serve as ecological indicators of the sand plain community, along with bear and post oaks, the holly, bearberry, and, occasionally, New England's only cactus, the prickly pear. It was not a forest upon which many colonists looked with favor. John Smith described the Cape as "onely a headland of high hils, over-growne with shrubby Pines, hurts and such trash." The Pilgrims spoke more favorably of both the Cape's forest and its soil—"excellent black earth" a spade's depth in thickness—but they eventually chose not to settle there. Those colonists who finally did establish settlements on the Cape encountered special problems.[19]

The pitch pine's most important adaptation to Cape Cod's dryness has to do with a phenomenon that might seem the scourge of a dry forest: fire. Natural and humanly induced forest fires have long been typical of the area. Driven by the Cape's strong southwestern summer winds, they have regularly destroyed species not adapted to their heat, and eliminated the humus layer of the soil so as to make the ground even drier than it already was. Although pitch pine is a highly flammable wood —the colonists used it for turpentine and preferred it for firewood—the tree possesses a dormant bud at the base of its trunk that allows it to sprout from its roots after the trunk has been destroyed, something few other conifers can do. Regular burning has thus guaranteed the maintenance of the pitch pine forest. In areas of the Cape protected from fire—lowland swamps, ponds, and the sheltered forests which the Pilgrims saw on the Province-town tip—the pitch pine could be replaced by moister forests containing large white oaks, white pine, an occasional hemlock, and the fire-sensitive beech.[20]

The effects of fire were by no means limited to Cape Cod; as we shall see, Indians made sure that they were very wide indeed. Throughout New England, fires which destroyed substantial portions of a hardwood forest created the conditions of full sunlight which species such as birch, white pine, and various shrubs needed in order to flourish. When Thomas Morton wrote of riding for ten miles through a forest in which there was "little or no other wood growing" but pine, he was probably describing the site of an old forest fire. Few forests so impressed the colonists as these old-burn stands of white pine, which contained what were easily the tallest trees in New England. The average height of a mature grove might be well over a hundred feet, with a few trees as much as five feet in diameter and 250 feet in height. The importance of the white pine to ship construction, especially for masts, made it one of the most sought after of colonial trees. "Of these," wrote Morton, "may be made rosin, pitch and tarre, which are such usefull commodities that if wee had them not from other Countries in Amity with England, our Navigation would decline." The effects of the English wood shortage led colonists to overemphasize the significance of pine in New England forests, thus obscuring the fact that the tree's chief habitats, other than old burned-over areas, were limited to dry ridge tops and sandy flood plains where it did not need to compete with other species for light. There was never the "infinite store" of it that Morton asserted.[21]

Not all the habitats of precolonial New England were forests. Some of the most important to both Indians and Europeans had no trees at all: whether rocky or sandy, the seashore was a zone of abundance from which both groups obtained food. Morton spoke of seeing oyster banks on Massachusetts Bay that were a mile in length. Wood declared that individual oysters could be as much as a foot long: once the animal was removed from its shell, it was so large "that it must admit of a division before you can well get it into your mouth." The movement of the tides brought thousands of lobsters into the shallow waters offshore, and exposed an "infinite store" of mussels and other shellfish. One observer described how a person running over exposed clam banks was soon "made all wet by their spouting of water," and said he had seen clams "as big as a penny white loaf" of English bread.[22]

As important as the shore itself were the salt marshes. Here the tides regularly flooded extensive inland areas with salt water, so that only two grasses—*Spartina patens* and *Spartina alterniflora*— were able to grow there. Because the grasses helped accumulate soil and so created a series of microenvironments from dry land to marsh to protected pools of water, they furnished a home for a wide variety of insects, fish, and birds. It was often in the salt marshes that the huge flocks of migratory waterfowl made their brief stops in Massachusetts Bay, creating those opportunities Wood spoke of to kill fifty "at a shot." But for the colonists the most striking thing about the marshes were the grasses themselves, which created the most extensive meadows to be found near the early settlements. All agreed with John Smith that the marshes contained "grasse plenty, though very long and thicke stalked, which being neither mowne nor eaten, is very ranke." The *Spartinas* had little of the sweetness of English grasses, and many colonists were dubious about their adequacy for hay—but they were often the only grass available. Wood warned prospective settlers that "hay ground is not in all places in New England," and suggested that those planning to keep cattle "choose the grassy valleys before the woody mountains." Inland, along the banks of rivers, colonists occasionally found rich grassy areas —called "intervals"—which served them well for hay, but all coastal settlements had to make sure of their access to the salt marshes.[23]

The precolonial landscape of New England was thus a patchwork. Even if one avoided exceptional areas like salt marshes or sand plains, one encountered tremendous variety even within the compass of a few square miles. The descent of a single hillside in southern New England, for instance, could easily carry one from a dry sunny forest of white and black oaks, white pine, and an occasional huckleberry or lowbush blueberry to a shaded valley buzzing with mosquitoes and containing red oak, tulip poplar, hemlock, and beech. In between might be chestnut and black birch, with the ubiquitous red maple appearing up and down the entire hillside.

Why a tree of a given species grew where it did was the result not only of ecological factors, such as climate, soil, and slope, but of history as well. A fire might shift a forest's composition from one group of species to another. A windstorm might blow over the mature trees of an entire tract of forest and allow the saplings growing beneath them to form a new canopy. Even a minor catastrophe, like the toppling of a single large tree, might create a microenvironment in the shadow of its uprooted base or in the sunlight of the newly broken canopy into which new species might move. Which species grew where in any particular place was thus the result of a cumulative sequence of ecological processes and historical events. The complexity of the precolonial ecosystem was one not merely of space but of time.[24]

The depth of that time was very great. The period during which Indians had inhabited the area had seen climatic warming transform southern New England from the glacial tundra of 12,500 years ago to a series of forests composed in turn of spruce, white pine, and finally, by about 7,000 years ago, the oaks and other hardwoods typical of the forest today. Because climatic trends involved such important overall shifts in forest composition, we tend to think of past forests in terms of the same generalized "vegetation zones" which supposedly existed at the coming of the Europeans. But such generalizations obscure too many details. In fact, the shifting composition of postglacial forests involved the complicated migrations not of homogeneous forest communities but of many individual species, each arriving by different routes and at different rates. At the same time that the supposed "spruce forest" dominated Connecticut, *Spartina* grasses were colonizing salt marshes. When "hardwood forests" burned several thousand years ago, they were replaced as they are today with stands of white pine, and these were replaced in turn when they aged and experienced windfall by a still different hardwood forest. Catastrophes—whether of fire, wind, or disease —continued to create drastic alterations of specific habitats even as general climatic trends were continuing. Just under 5,000 years ago, the region's hemlocks experienced an attack by some pathogenic organism that nearly destroyed them; it took over half a millennium for the population to recover. Events of this kind were not merely cyclical or self-equilibrating. They constitute a history of the ecosystem in which a unique linear sequence was imposed on the regularly recurring processes which ecology as a science seeks to describe.[25]

When human beings, Indian or European, inhabited and altered New England environments, they were a part of that linear history. Their activities often mimicked certain ecological processes that occurred in nature, but with a crucial difference. Whereas the natural ecosystem tended toward a patchwork of diverse communities arranged almost randomly on the landscape —its very continuity depending on that disorder—the human tendency was to systematize the patchwork and impose a more regular pattern on it. People sought to give their landscape a new purposefulness, often by simplifying its seemingly chaotic tangle.

Different peoples of course did this in different ways. Moreover, they chose different sets of habitats, different parts of the patchwork, to live in and reorder. When the Europeans first came to New England, they found a world which had been home to Indian peoples for over 10,000 years. But the way Indians had chosen to inhabit that world posed a paradox almost from the start for Europeans accustomed to other ways of interacting with the environment. Many European visitors were struck by what seemed to them the poverty of Indians who lived in the midst of a landscape endowed so astonishingly with abundance. As Thomas Morton wrote, "If this Land be not rich, then is the whole world poore." Here was a riddle: how could a land be so rich and its people so poor? At least in the eyes of many colonists, the Indians, blessed with such great natural wealth, nevertheless lived "like to our Beggers in England." To explain why this was so—or, alternatively, why the colonists perceived New England's earlier inhabitants in this way—we must turn to the Indians and their reasons for living as they did.[26]

3

SEASONS OF WANT

AND PLENTY

In describing New England's natural abundance so enthusiastically, the colonists were misleading in two ways; in the process, they revealed the assumptions by which they misconstrued the supposed "poverty" of the Indians. Those who sought to promote colonial enterprises tended to put the best possible face on everything they encountered in the New World. Selective reporting, exaggeration, and outright lies could all be useful tools in accomplishing this task. Captain Christopher Levett felt it necessary to inform readers of his 1628 account of New England that he would not "as some have done to my knowledge, speak more than is true." English readers must not be taken in by descriptions which made New England out to be a veritable paradise of milk and honey. "I will not tell you," Levett wrote,

> that you may smell the corn fields before you see the land; neither must men think that corn doth grow naturally, (or on trees,) nor will the deer come when they are called, or stand still and look on a man until he shoot him, not knowing a man from a beast; nor the fish leap into the kettle, nor

on the dry land, neither are they so plentiful, that you may dip them up in baskets, nor take cod in nets to make a voyage, which is no truer than that the fowls will present themselves to you with spits through them.

If the myths which Levett criticized had anything in common, it was their vision of a landscape in which wealth and sustenance could be achieved with little labor. Hopes for great windfall profits had fueled New World enterprises ever since the triumphs of Cortes, and were reinforced by traditions as old as the Garden of Eden. When English immigrants exaggerated the wealth of New England, they dreamed of a world in which returns to human labor were far greater than in England.[1]

Because their hopes led them to expect a land of plenty, early visitors introduced a second distortion into their accounts. Even when what they wrote was literally true, they often failed to note that it was not *always* true. Just as the habitats of New England formed a patchwork quilt on the landscape, the plenty of one being matched by the poverty of another, so too did those habitats change from month to month, the abundance of one season giving few clues to what a place might be like at other times of the year. Most early descriptions were written by spring and summer visitors, who naturally saw only the times when fish, fruit, and fowl were all too numerous to count. Would-be English settlers thus formed their vision of New England from accounts that concentrated the summer's seasonal wealth into an image of perpetual abundance. If the result was not disaster, it was at least disappointment. "When I remember the high commendations some have given of the place," wrote one chastened colonist, "I have thought the reason thereof to be this, that they wrote surely in strawberry time."[2]

New England's seasonal cycles were little different from those of Europe. If anything, its summers were hotter and its winters colder. Colonists were prevented from realizing this only by their own high expectations of laborless wealth: many initially seemed to believe that strawberry time would last all year. Captain Levett wrote of one early attempt at settlement in which the colonists "neither applied themselves to planting of corn nor taking of fish, more than for their present use, but went about to build castles in the air, and making of forts, neglecting the plenti-

ful time of fishing." They did so because their myths told them
that the plentiful times would never end, but their refusal to lay
up stores for the winter meant that many starved to death. The
pattern occurred repeatedly, whether at Sagadahoc, Plymouth,
or Massachusetts Bay: colonists came without adequate food supplies and died. At Plymouth alone, half the Pilgrims were dead
before the first winter was over. Those who had experienced the
New England cold knew better, and warned that new arrivals
who hoped to survive must bring provisions to last the year and
a half before settlements could become self-sustaining. "Trust
not too much on us for Corne at this time," wrote a spokesman
for the Pilgrims, "for by reason of this last company that came,
depending wholy upon us, we shall have little enough till harvest." This was hardly the advice one would send from a land of
infinite plenty. The problem was perhaps stated most plaintively
by the Massachusetts colonist John Pond, who in 1631 wrote his
parents, "I pray you remember me as your child . . . we do not
know how long we may subsist, for we cannot live here without
provisions from ould eingland."[3]

In New England, most colonists anticipated that they would
be able to live much as they had done in England, in an artisanal
and farming community with work rhythms, class relations, and
a social order similar to the one they had left behind—the only
difference being their own improved stature in society. There
were many misconceptions involved in this vision, but the one
most threatening to survival was the simple fact that establishing
European relations of production in the New World was a far
more complicated task than most colonists realized. Even to set
up farms was a struggle. Once colonists had done this, adjusting
to the New England ecosystem by re-creating the annual agricultural cycles which had sustained them in England, starving times
became relatively rare. But for the first year or two, before European subsistence patterns had been reproduced, colonists found
themselves forced to rely either on what little they had brought
with them or on what New England's inhabitants—whether English or Indian—were willing to provide. Few colonists expected
that they would have to go abegging like this. At most, they
contemplated supplementing their food stores by trading with
the Indians; and as one promoter argued, should the Indians be
reluctant to trade, it would be easy enough "to bring them all in

subjection, and make this provision." Many colonists arrived believing that they could survive until their first harvest simply by living as the Indians supposedly did, off the unplanted bounties of nature. Colonists were assured by some that Indian men got their livelihood with "small labour but great pleasure." Thomas Morton spoke of Indians for whom "the beasts of the forrest there doe serve to furnish them at any time when they please." If this were true, then surely Englishmen could do no worse. John Smith told his readers that, in New England, "nature and liberty affoords us that freely which in *England* we want, or it costeth us deerly." The willingness of colonists to believe such arguments, and hazard their lives upon them, was testimony to how little they understood both the New England environment and the ways Indians actually lived in it.[4]

A central fact of temperate ecosystems like those of New England is their periodicity: they are tied to overlapping cycles of light and dark, high and low tides, waxing and waning moons, and especially the long and short days which mean hot and cold seasons. Each plant and animal species makes its adjustments to these various cycles, so that the flowing of sap in trees, the migration of birds, the spawning of fish, the rutting of deer, and the fruiting of plants all have their special times of the year. A plant that stores most of its food energy in its roots during the winter will transfer much of that energy first to its leaves and then to its seeds as the warmer months progress. Such patterns of energy concentration are crucial to any creature which seeks to eat that plant. Because animals, including people, feed on plants and other animals, the ways they obtain their food are largely determined by the cycles in which other species lead their lives. Just as a fox's summer diet of fruit and insects shifts to rodents and birds during the winter, so too did the New England Indians seek to obtain their food wherever it was seasonally most concentrated in the New England ecosystem. Doing so required an intimate understanding of the habits and ecology of other species, and it was this knowledge that the English discovered they lacked.[5]

Indian communities had learned to exploit the seasonal diversity of their environment by practicing mobility: their communities characteristically refused to stay put. The principal social and economic grouping for precolonial New England Indians was the village, a small settlement with perhaps a few hundred inhabitants organized into extended kin networks. Villages, rather than the larger and better-known units called tribes or confederacies, were the centers around which Indian interactions with the environment revolved. But villages were not fixed geographical entities: their size and location changed on a seasonal basis, communities breaking up and reassembling as social and ecological needs required. Wherever villagers expected to find the greatest natural food supplies, there they went. When fish were spawning, many Indian families might gather at a single waterfall to create a dense temporary settlement in which feasting and celebration were the order of the day; when it was time to hunt in the fall, the same families might be found scattered over many square miles of land. All aspects of Indian life hinged on this mobility. Houses, consisting of wooden frames covered by grass mats or bark, were designed to be taken apart and moved in a few hours. For some groups, the shape of houses changed from season to season to accommodate different densities of population: small wigwams housing one or two families in the summer became in the winter extended longhouses holding many families. When food had to be stored while a village moved elsewhere, it was left in carefully constructed underground pit-barns, where it could be retrieved when needed. Tools and other property were either light and easily carried or just as readily abandoned and remade when needed in a new location. As Thomas Morton observed, "They love not to bee cumbered with many utensilles."[6]

The seasonal cycles within which a village moved depended on the habitats available to it: Indians who had access to the seashore, for instance, could lead rather different lives than their inland counterparts. Important as habitat differences were, however, the crucial distinction between Indian communities was whether or not they had adopted agriculture. In general, Indians south of the Kennebec River in Maine raised crops as part of their annual subsistence cycles; more northern Indians, on the other hand, as Verrazzano noted in 1524, showed "no sign of cultivation." Verrazzano quite reasonably attributed the absence of agriculture in the north to soil which would produce neither fruit nor grain "on account of its sterility": climatic conditions in fact made grain raising an increasingly risky business the farther north an Indian people lived. Because the ability to grow crops had drastic implications for the way a village conducted the rest of its food-gathering activities, it is best to begin our description of Indian subsistence strategies in the north, where Indians were entirely dependent on the natural abundance of the ecosystem. Only in the north did Indians live entirely as hunter-gatherers, people who bore at least superficial resemblance to the creatures of English fantasy who captured nature's bounties with "small labor but great pleasure."[7]

In the north, spring commenced "when the leaves begin to sprout, when the wild geese appear, when the fawns of moose attain to a certain size in the bellies of their mothers, and when the seals bear their young." Most especially, the northern spring began when the ice broke up; then inland populations moved to coastal sites where they repaired fishing gear—nets, tackle, weirs, birchbark canoes—in anticipation of the spawning runs. For Maine Indians who had access to the coast, probably well over half the yearly food supply came from the rivers and seashore. In late March, the smelt arrived in streams and rivers in such quantities that one could not put a "hand into the water, without encountering them." They were followed in April by the alewives, sturgeon, and salmon, so that spawning runs furnished a major share of the food supply from March through May. By early May, nonspawning fish were also providing food. Offshore were cod which had to be caught with hook and line. Closer to land were tidewater and ground fish, such as brook trout, smelt, striped bass, and flounder, all of which could be caught with weirs and nets, and the larger sturgeon and salmon, which were usually harpooned. In the tidal zone were the scallops, clams, mussels, and crabs which women and children gathered as a steady base for the village diet. As described by the Jesuit Pierre Biard, this phase of the northern Indians' subsistence cycle was especially flush: "From the month of May up to the middle of September, they are free from all anxiety about their food; for the cod are upon the coast, and all kinds of fish and shellfish."[8]

The arrival of the alewives also heralded the coming of the migratory birds, including the large ducks which Biard called bustards, whose eggs were over twice as large as ordinary European hens' eggs. Not only could women and children gather

birds' eggs while men fished; they could capture the birds themselves with snares or clubs. The Bird migrations made their biggest
contribution to Indian food supplies in April, May, September,
and October, when Canada geese, brants, mourning doves, and
miscellaneous ducks passed through; other birds, albeit in fewer
numbers, could be caught during the summer as well. By July
and August, strawberries, raspberries, and blueberries were ripening, providing food not only for Indians but for flocks of passenger pigeons and other birds which nested in the area. In
addition to birds, various coastal mammals—whales, porpoises,
walruses, and seals—were hunted and eaten. Nuts, berries, and
other wild plants were gathered as they became available. In all
ways, the summer was a time of plenty.

Things changed in September. Toward the middle of the
month, Indian populations moved inland to the smaller creeks,
where eels could be caught as they returned from their spawning
in the sea. From October through March, villages broke into
small family bands that subsisted on beaver, caribou, moose, deer,
and bear. Men were responsible for killing these animals, while
women maintained the campsite and did all hauling and processing of the slaughtered meat. If snows were heavy and animals
could be easily tracked, hunting provided an adequate food supply; if the snow failed to stay on the ground, on the other hand,
it was easy to starve. Northern Indians accepted as a matter of
course that the months of February and March, when the animals
they hunted were lean and relatively scarce, would be times of
little food.[9]

European visitors had trouble comprehending this Indian willingness to go hungry in the late winter months. They were
struck by the northern Indians' apparent refusal to store more
than a small amount of the summer's plenty for winter use. As
the Jesuit Chrétien Le Clercq remarked:

> They are convinced that fifteen to twenty lumps of meat,
> or of fish dried or cured in the smoke, are more than enough
> to support them for the space of five to six months. Since,
> however, they are a people of good appetite, they consume
> their provisions very much sooner than they expect. This
> exposes them often to the danger of dying from hunger,
> through lack of the provision which they could easily pos-

sess in abundance if they would only take the trouble to gather it.

Here again was the paradox of want in a land of plenty. To a European sensibility, it made no sense to go hungry if one knew in advance that there would be little food in winter. Colonists who starved did so because they learned too late how ill informed they had been about the New World's perpetual abundance. Although the myth died hard, those who survived it were reasonably quick to revise their expectations. When Europeans inquired why nonagricultural Indians did not do the same, the Indians replied, "It is all the same to us, we shall stand it well enough; we spend seven and eight days, even ten sometimes, without eating anything, yet we do not die." What they said was true: Indians died from starvation much less frequently than did early colonists, so there was a certain irony in European criticisms of Indians on this score. Whatever the contradictions of their own position, however, the colonists could not understand Indian attitudes toward winter food shortages. Consciously choosing hunger, rather than working harder in the leisurely times of summer, seemed a fool's decison.[10]

One effect of that choice, however, was to hold northern Indians to low population densities. The ecological principle known as Liebig's Law states that biological populations are limited not by the total annual resources available to them but by the minimum amount that can be found at the scarcest time of the year. Different species meet this restriction in different ways, and the mechanism—conscious or unconscious—whereby northern Indians restrained their fertility is not clear. However they accomplished this feat, its effects were self-evident: the low Indian populations of the precolonial northern forests had relatively little impact on the ecosystems they inhabited. The very abundance which so impressed the Europeans was testimony to this fact. By keeping population densities low, the food scarcities of winter guaranteed the abundance of spring, and contributed to the overall stability of human relationships to the ecosystem. In this, northern New England Indians were typical of hunting and gathering peoples around the world.[11]

The farming Indians of southern New England, among whom the earliest English colonists made their settlements, also engaged in hunting and gathering, but their ability to raise crops put them in a fundamentally different relationship with their environment. The very decision to engage in agriculture requires the creation of at least enough seed surplus to assure that planting can be done the following year, and opens the possibility of growing and storing enough food to carry a population through the winter with much less dependence on the vagaries of the hunt. Grain made up perhaps one-half to two-thirds of the southern New England diet, thereby reducing southern reliance on other foodstuffs; in comparison, northern Indians who raised no grain at all had to obtain two to three times more food energy from hunting and fishing. More importantly, nothing in the northern diet could be stored through the scarce times of winter as effectively as grain, making starvation a much less serious threat in the south than in the north.[12]

The ability of agriculture to smooth out the seasonal scarcities of wild foodstuffs had major consequences for the sizes of Indian populations in New England. The nonagricultural Indians of Maine sustained population densities, on average, of perhaps 41 persons per hundred square miles. The crop-raising Indians of southern New England, on the other hand, probably maintained 287 persons on an identical amount of land, a sevenfold difference. When these two broad groups were combined, the total Indian population of New England probably numbered somewhere between 70,000 and 100,000 people in 1600. (Lest this seem unimpressive, one should remember that the *English* population of New England was smaller than this even at the beginning of the eighteenth century, having reached only 93,000 people by 1700.) The crucial role of agriculture in maintaining so large an Indian population in precolonial New England is clear: although agricultural and nonagricultural peoples inhabited roughly equal areas of southern and northern New England respectively, those who raised crops contributed over 80 percent of the total population.[13]

Although southern Indians engaged in many of the same annual hunting and fishing activities as northern ones, their concentration on the raising of crops can be seen even in the names they gave their months. Northern Indians named their lunar months in terms of seasonal changes in animal populations, referring to the egg laying of birds, the running of salmon, the molting of geese, the hibernation of bears, and so on. By contrast, southern Indians chose the names of their months with an entirely different emphasis. The fur trader John Pynchon recorded that the Agawam Indian village near Springfield, Massachusetts, began its year with the month of Squannikesos, which included part of April and part of May, and whose name meant "when they set Indian corn." This was followed by various months whose names indicated the weeding of corn, the hilling of corn, the ripening of corn, the coming of the frost, the middle of winter, the thawing of ice, and the catching of fish. The southern cycle of months was thus remarkable in having only a single reference to the animals which so dominated the northern calendar, an indication of how much agriculture had transformed Indian lives there.[14]

As the Agawam calendar shows, southern Indians began their annual subsistence cycles by moving to their summer fields and preparing the ground by working it with clamshell hoes. According to the Dutch traveler Isaack de Rasieres, the Indians "make heaps like molehills, each about two and a half feet from the others, which they sow or plant in April with maize, in each heap five or six grains." Because the earth was not stirred deeply by this method, much of the soil was left intact and erosion was thereby held to a minimum. As the young plants grew, soil was raised around them to create low mounds which strengthened their roots against the attacks of birds. Maize was not an easy crop to raise: as de Rasieres noted, it was "a grain to which much labor must be given, with weeding and earthing-up, or it does not thrive." Perhaps partly for this reason, Indian farmers, unlike European ones, used their cornfields to raise more than just corn. When Champlain observed Indian fields near the mouth of the Saco River, he noted that

> with the corn they put in each hill three or four Brazilian beans [kidney beans], which are of different colors. When they grow up, they interlace with the corn, which reaches to the height of from five to six feet; and they keep the ground very free from weeds. We saw there many squashes, and pumpkins, and tobacco, which they likewise cultivate.

It was not an agriculture that looked very orderly to a European eye accustomed to monocultural fields. Cornstalks served as beanpoles, squashes sent their tendrils everywhere, and the entire surface of the field became a dense tangle of food plants. But, orderly or not, such gardens had the effect, as John Winthrop, Jr., said, of "loading the Ground with as much as it will beare," creating very high yields per acre, discouraging weed growth, and preserving soil moisture. Moreover, although Indians may or may not have realized it, the resulting harvest of beans and corn provided the amino acids necessary for a balanced diet of vegetable protein.[15]

Except for tobacco, crops were primarily the responsibility of women. Roger Williams wrote that Indian women "constantly beat all their corne with hand: they plant it, dresse it, gather it, barne it, beat it, and take as much paines as any people in the world" with it. As with the hunting Indians of northern New England, the sexual division of labor for the agricultural peoples of southern New England was very well defined, women performing those jobs which were most compatible with simultaneous child-care. This meant tasks which were generally repetitive, which could be easily interrupted, which did not require travel too far from home, and which did not suffer if one performed them while giving most of one's attention to the children. In the nonagricultural north, women's work involved gathering shellfish and birds on the shore, collecting wild plants, trapping small rodents, making garments, keeping camp, and the whole range of food-processing activities; but meat gathered by men probably supplied half or more of a village's food. In the south, on the other hand, agriculture changed this sexual division and made women much more important than men in providing food. A single Indian woman could raise anywhere from twenty-five to sixty bushels of corn by working an acre or two, enough to provide half or more of the annual caloric requirements for a family of five. When corn was combined with the other foods for which they were responsible, women may have contributed as much as three-fourths of a family's total subsistence needs.[16]

Crops were planted between March and late June, the event often being timed by the leafing of certain trees or the arrival of the alewives. While women worked the fields, men erected weirs on the rivers and fished the spring spawning runs. By March, most beans and corn remaining from the previous harvest were probably needed as seed for planting, so that fish and migratory birds became the chief sources of food from late winter through midsummer. Contrary to what American myth has long held, it is quite unlikely that alewives or other fish were used as fertilizer in Indian fields, notwithstanding the legendary role of the Pilgrims' friend Squanto in teaching colonists this practice. Squanto probably learned the technique while being held captive in Europe, and if any Indians used it in New England, they did so in an extremely limited area. Having no easy way to transport large quantities of fish from river to field, and preferring quite sensibly to avoid such back-breaking work, Indians simply abandoned their fields when the soil lost its fertility. As William Wood wrote, "The Indians who are too lazy to catch fish plant corn eight or ten years in one place without it, having very good crops." Fertilizing fields with fish, as the English eventually did, seemed to Indians a wholly unnecessary labor.[17]

Once crops were planted and weeded, they needed less attention for two or three months, until the ripening corn had to be guarded against marauding birds before being harvested. (De Rasieres explained how some birds, probably passenger pigeons, were known as "maize thieves" because "they flatten the corn in any place where they alight, just as if cattle had lain there.") During these months, villages tended to disperse and families moved their individual wigwams to other planting and gathering sites. Women, who owned the wigwams and most household goods, moved their camps from field to field as necessary, and then to points along the coast where they gathered seafood and the cattails used in making mats for wigwams. Camps occasionally had to be moved in the summer simply to escape the fleas which tended to breed around human habitations. Wigwams were also moved if a death occurred in one, or if a settlement was threatened by war.[18]

Men fanned out from these bases for extended fishing and hunting trips. They might disappear into the woods for ten days at a time to build a dugout canoe that would allow them to fish deep water with harpoon or hook and line. Southern New England boats were made from decay-resistant chestnut and were heavy enough to require several hands to launch; in the north, paper birch, which did not grow in southeastern New England, was used to create the much lighter and more familiar birchbark canoes. Whether birch or chestnut, these tippy boats might be

taken a mile or more offshore at night to hunt sturgeon by torch-
light, or be run down the rapids of rivers in search of salmon or
eels. Used for these purposes, canoes could be very dangerous
indeed. Roger Williams spoke from personal experience when he
said, "It is wonderfull to see how they will venture in those
Canoes, and how (being oft overset as I have myself been with
them) they will swim a mile, yea two or more safe to Land." Such
danger was typical of male work. Whereas the relatively steady
labor of agriculture and gathering allowed women to provide the
largest share of a village's food without moving far from home,
the hunting and fishing of animal protein had much different
requirements. These activities took men far from the main camp
for many days at a time, and exposed them to much greater risk
of injury or death. Hunting and fishing both had irregular work
rhythms which sometimes required many intense hours of labor
under hard conditions, and sometimes long hours of idleness.
Times in camp were often periods of relative leisure and recuper-
ation for men.[19]

As summer drew to a close, female food production reached a
climax and male hunting activities began to contribute a greater
share of the village's food. Autumn saw the harvesting of corn in
addition to the gathering of acorns, chestnuts, groundnuts, and
other wild plants. It was a time of extensive festivals when many
hundreds of people gathered in dense settlements and consumed
much of this surplus food. Gambling, dancing, and eating were
combined with rituals—similar to the potlatch ceremonies of the
Pacific Northwest—in which wealthy individuals gave away
much of what they owned to establish reciprocal relations of
obligation with potential followers or allies. The harvest saw
greater surplus than any other time of year, and so was often the
preferred season for going to war, when food stores both at home
and in enemy territory would be at their peak. But once the
harvest celebrations were over, Indian households struck their
wigwams, stored the bulk of their corn and beans, and moved to
campsites to conduct the fall hunt.[20]

From October to December, when animals like bear and deer
were at their fattest, southern villages, much like their counterparts in the north, broke into small bands to assure maximum
coverage of the hunting territory. Again the sexual division of
labor came into play. Men hunted steadily, using a variety of

techniques. Game might be stalked with bow and arrow by a lone hunter or by groups of two or three hundred men working together. It might be snared with traps specially designed to capture a single species; William Bradford, for instance, accidentally walked into a trap strong enough to hold a full-grown deer. Or game might be run between specially planted hedges more than a mile in length until it was finally driven onto the weapons of waiting hunters. Nothing required a greater knowledge of animal behavior than the winter hunt. While men remained in the field, women hauled dead game back to camp. There they butchered and processed it, preparing the hides for clothing, cooking the meat, and smoking some of it for use later in the winter.[21]

By late December, when the snows finally came, the village had probably reassembled in heavily wooded valleys well protected from the weather, where fuel for campfires was easy to obtain. For the rest of the winter, men continued to hunt and fish the surrounding area on snowshoes, while women remained in camp making garments and living on meat and stored grain. Especially for men away from camp, winter was a time of occasional hunger between kills; most carried only a small store of parched corn flour called *nocake* as traveling fare. Like their hunting kindred to the north, they accepted such hunger as inevitable and bore it with stoicism. As Samuel Lee reported, the Indians were "very patient in fasting, & will gird in their bellies till they meet with food; but then none more gluttons or drunk on occasion. Theyle eat 10 times in 24 houres, when they have a beare or a deere."[22]

The hunt provided a crucial source of protein and vitamins during the winter. A single season's catch for a southern New England village of about 400 inhabitants might bring in over 8,500 pounds of edible deer meat and over 7,000 pounds of bear, the two animals which together contributed more than three-fourths of an inland village's winter meat supply. (Coastal Indians who relied more heavily on seafood killed smaller amounts of large game.) Whether or not this meat was essential to a community's survival—given the availability of stored beans and grain—the skins of these and other furbearing animals would furnish the village's clothing for the following year. Simple measurements of caloric content thus tend to undervalue the importance of the fall and winter hunt to an agricultural village's subsistence cycle. Hundreds of square miles had to be stalked to obtain skins for the skirts, leggings, shirts, moccasins, and other articles of clothing Indians would need in the months ahead.[23]

The relationship of the southern New England Indians to their environment was thus, if anything, even more complicated than that of the northern Indians. To the seasons of hunting and fishing shared by both groups were added the agricultural cycles which increased the available food surplus and so enabled denser populations to sustain themselves. In both areas, the mobility of village sites and the shift between various subsistence bases reduced potential strains on any particular segment of the ecosystem, keeping the overall human burden low. But in clearing land for planting and thus concentrating the food base, southern Indians were taking a most important step in reshaping and manipulating the ecosystem.

Clearing fields was relatively easy. By setting fire to wood piled around the base of standing trees, Indian women destroyed the bark and so killed the trees; the women could then plant corn amid the leafless skeletons that were left. During the next several years, many of the trees would topple and could be entirely removed by burning. As one Indian remembered, "An industrious woman, when great many dry logs are fallen, could burn off as many logs in one day as a smart man can chop in two or three days time with an axe." However efficient they were at such clearing, Indian women were frugal with their own labor, and sought to avoid even this much work for as long as they could. That meant returning to the same field site for as long as possible, usually eight to ten years. In time, the soil gradually lost its fertility and eventually necessitated movement to a new field. (Soil exhaustion was to some extent delayed by the action of the nitrogen-fixing beans which Indian women planted with the corn; whether they were aware of it or not, this was one of the side benefits of planting multicrop fields.)[24]

The annual reoccupation of fixed village and planting sites meant that the area around field and camp experienced heavy human use: intensive food gathering, the accumulation of garbage, and, most importantly, the consumption of firewood. One of the main reasons Indians moved to winter camps was that their summer sites had been stripped of the fuel essential for winter fires. Indians believed in big fires—one colonist said that "their Fire is instead of our bed cloaths"—and burned wood heavily all night long, both summer and winter. Such practices could not long be maintained on a single site. As Morton said, "They use not to winter and summer in one place, for that would be a reason to make fuell scarce." The Indians were thus no strangers to the fuel shortages so familiar to the English, even if Indian scarcities were more local. When Verrazzano found twenty-five to thirty leagues of treeless land in Narragansett Bay, or Higginson spoke of thousands of acres in a similar state near Boston, they were observing the effects of agricultural Indians returning to fixed village sites and so consuming their forest energy supply. Indeed, when the Indians wondered why English colonists were coming to their land, the first explanation that occurred to them was a fuel shortage. Roger Williams recounted:

> This question they oft put to me: Why come the *Englishmen* hither? and measuring others by themselves; they say, It is because you want *firing*: for they, having burnt up the *wood* in one place, (wanting draughts [animals] to bring *wood* to them) they are faine to follow the *wood*; and so to remove to a fresh new place for the *woods* sake.

Williams regarded this merely as a quaint instance of Indian provincialism, but in one ironic sense, given what we know of the English forests of the seventeenth century, the Indians were perhaps shrewder than he knew.[25]

The effect of southern New England Indian villages on their environment was not limited to clearing fields or stripping forests for firewood. What most impressed English visitors was the Indians' burning of extensive sections of the surrounding forest once or twice a year. "The Salvages," wrote Thomas Morton, "are accustomed to set fire of the Country in all places where they come, and to burne it twize a yeare, viz: at the Spring, and the fall of the leafe." Here was the reason that the southern forests were so open and parklike; not because the trees naturally grew thus, but because the Indians preferred them so. As William Wood observed, the fire "consumes all the underwood and rubbish which otherwise would overgrow the country, making it unpassable, and spoil their much affected hunting." The result was a forest of large, widely spaced trees, few shrubs, and much grass and herbage. "In those places where the Indians inhabit," said Wood, "there is scarce a bush or bramble or any cumbersome underwood to be seen in the more champion ground." By removing underwood and fallen trees, the Indians reduced the total accumulated fuel at ground level. With only small nonwoody plants to consume, the annual fires moved quickly, burned with relatively low temperatures, and soon extinguished themselves. They were more ground fires than forest fires, not usually involving larger trees, and so they rarely grew out of control. Fires of this kind could be used to drive game for hunting, to clear fields for planting, and, on at least one occasion, to fend off European invaders.[26]

Northern Indians do not appear to have engaged in such burning. Because they did not practice agriculture and so were less tied to particular sites, they had less incentive to alter the environment of a given spot. Their chief mode of transportation was the canoe, so that they had less need of an open forest for traveling. Moreover, many of the northern tree species were not well adapted to repeated burning, and northern forests tended to accumulate enough fuel at ground level that, once a fire got started, it usually reached the canopy and burned out of control. Conditions in southern New England were quite different. Denser, fixed settlements encouraged heavy use of more limited forest areas, and most inland travel was by land. The trees of the southern forest, once fully grown, suffered little more than charred bark if subjected to ground fires of short duration. If destroyed, they regenerated themselves by sprouting from their roots: chestnuts, oaks, and hickories, the chief constituents of the southern upland forests, are in fact sometimes known as "sprout hardwoods." Repeated fires tended to destroy trees and shrubs which lacked this ability, including hemlock, beech, and juniper. Even the white pine, which often sprang up after large forest fires, tended be killed off if subjected to regular burning because of its inability to sprout, and so was uncommon in the vicinity of active Indian settlements.[27]

Colonial observers understood burning as being part of Indian efforts to simplify hunting and facilitate travel; most failed to see its subtler ecological effects. In the first place, it increased the rate at which forest nutrients were recycled into the soil, so that grasses, shrubs, and nonwoody plants tended to grow more luxuriantly following a fire than they had before. Especially on old

Indian fields, fire created conditions favorable to strawberries, blackberries, raspberries, and other gatherable foods. Grasses like the little bluestem were rare in a mature forest, but in a forest burned by Indians they became abundant. The thinning of the forest canopy, which resulted from the elimination of smaller trees, allowed more light to reach the forest floor and further aided such growth. The soil became warmer and drier, dis-couraging tree species which preferred moister conditions—beech, sugar maple, red maple, black birch—and favoring drier species like oaks when regular burning was allowed to lapse. Burning also tended to destroy plant diseases and pests, not to mention the fleas which inevitably became abundant around In-dian settlements. Roger Williams summed up these effects by commenting that "this burning of the Wood to them they count a Benefit, both for destroying of vermin, and keeping downe the Weeds and thickets."[28]

Selective Indian burning thus promoted the mosaic quality of New England ecosystems, creating forests in many different states of ecological succession. In particular, regular fires pro-moted what ecologists call the "edge effect." By encouraging the growth of extensive regions which resembled the boundary areas between forests and grasslands, Indians created ideal habitats for a host of wildlife species. Of all early American observers, only the astute Timothy Dwight seems to have commented on this phenomenon. "The object of these conflagrations," he wrote, "was to produce fresh and sweet pasture for the purpose of allur-ing the deer to the spots on which they had been kindled." The effect was even subtler than Dwight realized: because the en-larged edge areas actually raised the total herbivorous food sup-ply, they not merely attracted game but helped create much larger populations of it. Indian burning promoted the increase of exactly those species whose abundance so impressed English colonists: elk, deer, beaver, hare, porcupine, turkey, quail, ruffed grouse, and so on. When these populations increased, so did the carnivorous eagles, hawks, lynxes, foxes, and wolves. In short, Indians who hunted game animals were not just taking the "un-planted bounties of nature"; in an important sense, they were harvesting a foodstuff which they had consciously been instru-mental in creating.[29]

Few English observers could have realized this. People accustomed to keeping domesticated animals lacked the conceptual tools to realize that Indians were practicing a more distant kind of husbandry of their own. To the colonists, only Indian women appeared to do legitimate work; the men idled away their time in hunting, fishing, and wantonly burning the woods, none of which seemed like genuinely productive activities to Europeans. English observers often commented about how hard Indian women worked. "It is almost incredible," Williams wrote, "what burthens the poore women carry of *Corne*, of *Fish*, of *Beanes*, of *Mats*, and a childe besides." The criticism of Indian males in such remarks was usually explicit. "Their wives are their slaves," wrote Christopher Levett, "and do all the work; the men will do nothing but kill beasts, fish, etc." For their part, Indian men seemed to acknowledge that their wives were a principal source of wealth and mocked Englishmen for not working their wives harder. According to the lawyer Thomas Lechford, "They say, *Englishman* much foole, for spoiling good working creatures, meaning women: And when they see any of our *English* women sewing with their needles, or working coifes, or such things, they will cry out, Lazie squaes."[30]

Part of the problem with these cross-cultural criticisms was the inability or refusal by either side to observe fully how much each sex was contributing to the total food supply. Indian men, seeing Englishmen working in the fields, could not understand why English *women* were not doing such work. At the same time, they failed to see the contributions colonial women were actually making: gardening, cooking, spinning and weaving textiles, sewing clothing, tending milch cows, making butter and cheese, caring for children, and so on. The English, for their part, had trouble seeing hunting and fishing—which most regarded as leisure activities—as involving real labor, and so tended to brand Indian men as lazy. "The Men," wrote Francis Higginson, "for the most part live idely, they doe nothing but hunt and fish: their wives set their Corne and doe all their other worke." It is quite possible that Indian women—like women in many cultures—did indeed bear a disproportionate share of the work burden. But even if the advent of agriculture in southern New England had shifted the balance between meat and vegetables in the Indian diet—lowering the importance of meat and incidentally changing the significance of each sex's role in acquiring food—the annual subsistence cycle still saw Indian communities giving considerable attention to hunting meat, the traditionally more masculine activity. As we shall see, the English used this Indian reliance on hunting not only to condemn Indian men as lazy savages but to deny that Indians had a rightful claim to the land they hunted. European perceptions of what constituted a proper use of the environment thus reinforced what became a European ideology of conquest.[31]

The relationships of the New England Indians to their environment, whether in the north or the south, revolved around the wheel of the seasons: throughout New England, Indians held their demands on the ecosystem to a minimum by moving their settlements from habitat to habitat. As one of the earliest European visitors noted, "They move . . . from one place to another according to the richness of the site and the season." By using other species when they were most plentiful, Indians made sure that no single species became overused. It was a way of life to match the patchwork of the landscape. On the coast were fish and shellfish, and in the salt marshes were migratory birds. In the forests and lowland thickets were deer and beaver; in cleared upland fields were corn and beans; and everywhere were the wild plants whose uses were too numerous to catalog. For New England Indians, ecological diversity, whether natural or artificial, meant abundance, stability, and a regular supply of the things that kept them alive.[32]

The ecological relationships which the English sought to reproduce in New England were no less cyclical than those of the Indians; they were only simpler and more concentrated. The English too had their seasons of want and plenty, and rapidly adjusted their false expectations of perpetual natural wealth to match New World realities. But whereas Indian villages moved from habitat to habitat to find maximum abundance through minimal work, and so reduce their impact on the land, the English believed in and required permanent settlements. Once a village was established, its improvements—cleared fields, pastures, buildings, fences, and so on—were regarded as more or less fixed features of the landscape. English fixity sought to replace Indian mobility; here was the central conflict in the ways Indians and colonists interacted with their environments. The struggle was over two ways of living and using the seasons of the year, and it expressed itself in how two peoples conceived of property, wealth, and boundaries on the landscape.

4

BOUNDING THE LAND

To take advantage of their land's diversity, Indian villages had to be mobile. This was not difficult as long as a family owned nothing that could not be either stored or transported on a man's or —more probably—a woman's back. Clothing, baskets, fishing equipment, a few tools, mats for wigwams, some corn, beans, and smoked meat: these constituted most of the possessions that individual Indian families maintained during their seasonal migrations. Even in southern New England, where agriculture created larger accumulations of food than existed among the hunter-gatherer peoples of the north, much of the harvest was stored in underground pits to await later visits and was not transported in large quantities. The need for diversity and mobility led New England Indians to avoid acquiring much surplus property, confident as they were that their mobility and skill would supply any need that arose.

This, then, was a solution to the riddle Thomas Morton had posed his European readers. If English visitors to New England thought it a paradox that Indians seemed to live like paupers in a landscape of great natural wealth, then the problem lay with

English eyesight rather than with any real Indian poverty. To those who compared Massachusetts Indians to English beggars, Morton replied, "If our beggers of England should, with so much ease as they, furnish themselves with foode at all seasons, there would not be so many starved in the streets." Indians only *seemed* impoverished, since they were in fact "supplied with all manner of needefull things, for the maintenance of life and lifelyhood." Indeed, said Morton, the leisurely abundance of Indian life suggested that there might be something wrong with *European* no tions of wealth: perhaps the English did not know true riches when they saw them. In a passage undoubtedly intended to infuriate his Puritan persecutors, Morton counterposed to the riddle of Indian poverty a riddle of Indian wealth: "Now since it is but foode and rayment that men that live needeth (though not all alike,) why should not the Natives of New England be sayd to live richly, having no want of either?"[1]

Why not indeed? It was not a question that sat well with the New England Puritans, who had banished Morton for just such irreverence (not to mention his rival trade with the Indians). Criticism of Indian ways of life was a near-constant element in early colonial writing, and in that criticism we may discover much about how colonists believed land should be used. "The *Indians,*" wrote Francis Higginson, "are not able to make use of the one fourth part of the Land, neither have they any setled places, as Townes to dwell in, nor any ground as they challenge for their owne possession, but change their habitation from place to place." A people who moved so much and worked so little did not deserve to lay claim to the land they inhabited. Their supposed failure to "improve" that land was a token not of their chosen way of life but of their laziness. "Much might they benefit themselves," fumed William Wood, "if they were not strong fettered in the chains of idleness; so as that they had rather starve than work, following no employments saving such as are sweetened with more pleasures and profit than pains or care." Few Indians, of course, had actually starved in precolonial times, so Wood's criticism boiled down to an odd tirade against Indians who chose to subsist by labor they found more pleasurable than hateful. (Ironically, this was exactly the kind of life that at least some colonists fantasized for themselves in their visions of the natural bounty of the New World.) Only the crop-planting (and therefore supposedly overworked) women were exempted from such attacks. As we have seen, the full scorn of English criticism was reserved for Indian males, whose lives were perhaps too close to certain English pastoral and aristocratic fantasies for Calvinists to tolerate. At a time when the royalist Izaak Walton would soon proclaim the virtues of angling and hunting as pastimes, the Puritan objections to these "leisure" activities carried political as well as moral overtones.[2]

More importantly, English colonists could use Indian hunting and gathering as a justification for expropriating Indian land. To European eyes, Indians appeared to squander the resources that were available to them. Indian poverty was the result of Indian waste: underused land, underused natural abundance, underused human labor. In his tract defending "the Lawfulness of Removing Out of England into the Parts of America," the Pilgrim apologist Robert Cushman argued that the Indians were "not industrious, neither have art, science, skill or faculty to use either the land or the commodities of it; but all spoils, rots, and is marred for want of manuring, gathering, ordering, etc." Because the Indians were so few, and "do but run over the grass, as do also the foxes and wild beasts," Cushman declared their land to be "spacious and void," free for English taking.[3]

Colonial theorists like John Winthrop posited two ways of owning land, one natural and one civil. Natural right to the soil had existed "when men held the earth in common every man sowing and feeding where he pleased." This natural ownership had been superseded when individuals began to raise crops, keep cattle, and improve the land by enclosing it; from such actions, Winthrop said, came a superior, civil right of ownership. That these notions of land tenure were ideological and inherently Eurocentric was obvious from the way Winthrop used them: "As for the Natives in New England," he wrote, "they inclose noe Land, neither have any setled habytation, nor any tame Cattle to improve the Land by, and soe have noe other but a Naturall Right to those Countries." By this argument, only the fields planted by Indian women could be claimed as property, with the happy result, as Winthrop said, that "the rest of the country lay open to any that could and would improve it." The land was a *vacuum Domicilium* waiting to be inhabited by a more productive people. "In a vacant soyle," wrote the minister John Cotton, "hee that taketh possession of it, and bestoweth culture and husbandry upon it, his Right it is."[4]

This was, of course, little more than an ideology of conquest conveniently available to justify the occupation of another people's lands. Colonists occasionally admitted as much when they needed to defend their right to lands originally purchased from Indians: in order for Indians legitimately to sell their lands, they had first to own them. Roger Williams, in trying to protect Salem's claim to territory obtained from Indians rather than from the English Crown, argued that the King had committed an "injustice, in giving the Countrey to his *English* Subjects, which belonged to the Native *Indians.*" Even if the Indians used their land differently than did the English, Williams said, they nevertheless possessed it by right of first occupancy and by right of the ecological changes they had wrought in it. Whether or not the Indians conducted agriculture, they "hunted all the Countrey over, and for the expedition of their hunting voyages, they burnt up all the underwoods in the Countrey, once or twice a yeare." Burning the woods, according to Williams, was an improvement that gave the Indians as much right to the soil as the King of England could claim to the royal forests. If the English could invade Indian hunting grounds and claim right of ownership over them because they were unimproved, then the Indians could do likewise in the royal game parks.[5]

It was a fair argument. Williams's opponents could only reply that English game parks were not just hunted but also used for cutting timber and raising cattle; besides, they said, the English King (along with lesser nobles holding such lands) performed other services for the Commonwealth, services which justified his large unpeopled holdings. If these assertions seemed a little lame, designed mainly to refute the technical details of Williams's argument, that was because the core of the dispute lay elsewhere. Few Europeans were willing to recognize that the ways Indians inhabited New England ecosystems were as legitimate as the ways Europeans *intended* to inhabit them. Colonists thus rationalized their conquest of New England: by refusing to extend the rights of property to the Indians, they both trivialized the ecology of Indian life and paved the way for destroying it. "We did not conceive," said Williams's opponents with fine irony, "that it is a just Title to so vast a Continent, to make no other improvement of millions of Acres in it, but onely to burn it up for pastime."[6]

Whether denying or defending Indian rights of land tenure, most English colonists displayed a remarkable indifference to what the Indians themselves thought about the matter. As a result, we have very little direct evidence in colonial records of the New England Indians' conceptions of property. To try to reconstruct these, we must use not only the few early fragments available to us but a variety of evidence drawn from the larger ethnographic literature. Here we must be careful about what we mean by "property," lest we fall into the traps English colonists have set for us. Although ordinary language seems to suggest that property is generally a simple relationship between an individual person and a thing, it is actually a far more complicated social institution which varies widely between cultures. Saying that A owns B is in fact meaningless until the society in which A lives agrees to allow A a certain bundle of rights over B and to impose sanctions against the violation of those rights by anyone else. The classic definition is that of Huntington Cairns: "the property relation is triadic: 'A owns B against C,' where C represents all other individuals." Unless the people I live with recognize that I own something and so give me certain unique claims over it, I do not possess it in any meaningful sense. Moreover, different groups will permit me different bundles of rights over the same object. To define property is thus to represent boundaries between people; equally, it is to articulate at least one set of conscious ecological boundaries between people and things.[7]

This suggests that there are really two issues involved in the problem of Indian property rights. One is individual *ownership*, the way the inhabitants of a particular village conceived of property vis-à-vis each other; and the other is collective *sovereignty*, how everyone in a village conceived of their territory (and political community) vis-à-vis other villages. An individual's or a family's rights to property were defined by the community which recognized those rights, whereas the community's territorial claims were made in opposition to those of other sovereign groups. Distinctions here can inevitably become somewhat artificial. Because kin networks might also have territorial claims—both *within* and *across* villages—even the village is sometimes an arbitrary unit in which to analyze property rights: ownership and sovereignty among Indian peoples could shade into each other in a way Europeans had trouble understanding. For this reason, the nature of Indian political communities is crucial to any discussion of property rights.

A village's right to the territory which it used during the various seasons of the year had to be at least tacitly accepted by other villages or, if not, defended against them. Territorial rights of this kind, which were expressions of the entire group's collective right, tended to be vested in the person of the sachem, the leader in whom the village's political identity at least symbolically inhered. Early English visitors who encountered village sachems tended to exaggerate their authority by comparing them to European kings: Roger Williams and John Josselyn both baldly asserted of New England Indians that "their Government is Monarchicall." Comparison might more aptly have been made to the relations between lords and retainers in the early Middle Ages of Europe. In reality, sachems derived their power in many ways: by personal assertiveness; by marrying (if male) several wives to proliferate wealth and kin obligations; by the reciprocal exchange of gifts with followers; and, especially in southern New England, by inheriting it from close kin. Although early documents are silent on this score, kin relations undoubtedly cemented networks both of economic exchange and of political obligation, and it was on these rather than more formal state institutions that sachems based their authority. As William Wood remarked, "The kings have not many laws to command by, nor have they any annual revenues."[8]

Polity had less the abstract character of a monarchy, a country, or even a tribe, than of a relatively fluid set of personal relationships. Although those relationships bore some resemblance to the dynastic politics of early modern Europe—a resemblance several historians have recently emphasized—they were crucially different in not being articulated within a state system. Kinship and personality rather than any alternative institutional structure organized power in Indian communities. Both within and between villages, elaborate kin networks endowed individuals with greater or lesser degrees of power. A sachem—who could be either male or female—asserted authority only in consultation with other powerful individuals in the village. Moreover, the sachem of one village might regularly pay tribute to the sachem of another, thus acknowledging a loose hierarchy between villages and sachems. Such hierarchies might be practically unimportant until some major conflict or external threat arose, whereupon the communities assembled into a larger confederacy until the problem was solved. The result, like Indian subsistence patterns, entailed a good deal more flexibility and movement than Europeans were accustomed to in their political institutions. As the missionary Daniel Gookin indicated, it was a very shifting politics:

> Their sachems have not their men in such subjection, but that very frequently their men will leave them upon distaste or harsh dealing, and go and live among other sachems that can protect them: so that their princes endeavour to carry it obligingly and lovingly unto their people, lest they should desert them, and thereby their strength, power, and tribute would be diminished.[9]

Insofar as a village "owned" the land it inhabited, its property was expressed in the sovereignty of the sachem. "Every sachem," wrote Edward Winslow, "knoweth how far the bounds and limits of his own Country extendeth." For all of their differences, a sachem "owned" territory in a manner somewhat analogous to the way a European monarch "owned" an entire European nation: less as personal real estate than as the symbolic possession of a whole people. A sachem's land was coterminous with the area within which a village's economic subsistence and political sanctions were most immediately expressed. In this sovereign sense, villages were fairly precise about drawing boundaries among their respective territories. When Roger Williams wrote that "the *Natives* are very exact and punctuall in the bounds of their Lands, belonging to this or that Prince or People," he was refuting those who sought to deny that legitimate Indian property rights existed. But the rights of which he spoke were not ones of individual ownership; rather, they were sovereign rights that defined a village's political and ecological territory.[10]

The distinction becomes important in the context of how such territorial rights could be alienated. Williams said that he had "knowne them make bargaine and sale amongst themselves for a small piece, or quantity of Ground," suggesting that Indians were little different from Europeans in their sense of how land could be bought or sold. When two sachems made an agreement to transfer land, however, they did so on behalf of their two political or kinship communities, as a way of determining the customary rights each village would be allowed in a given area. An instructive example of this is the way Roger Williams had to correct John Winthrop's confusion over two islands which Winthrop thought Williams had bought from the Narragansett sachem Miantonomo. Williams had indeed gotten permission to use the islands for grazing hogs—a land transaction of sorts had taken place—but it was emphatically not a purchase. "Be pleased to understand," cautioned Williams, "your great mistake: neither of them were sold properly, for a thousand fathom [of wampum] would not have bought either, by strangers. The truth is, not a penny was demanded for either, and what was paid was only gratuity, though I choose, for better assurance and form, to call it sale." What had been transacted, as Williams clearly understood, was more a diplomatic exchange than an economic one. Miantonomo, like other New England sachems, had no intention of conducting a market in real estate.[11]

That this was so can best be seen by examining how a village's inhabitants conceived of property *within* its territory. Beginning with personal goods, ownership rights were clear: people owned what they made with their own hands. Given the division of labor, the two sexes probably tended to possess the goods that were most closely associated with their respective tasks: women owned baskets, mats, kettles, hoes, and so on, while men owned bows, arrows, hatchets, fishing nets, canoes, and other hunting tools. But even in the case of personal goods, there was little sense either of accumulation or of exclusive use. Goods were owned because they were useful, and if they ceased to be so, or were needed by someone else, they could easily be given away. "Although every proprietor knowes his own," said Thomas Morton, "yet all things, (so long as they will last), are used in common amongst them." Not surprisingly, theft was uncommon in such a world.[12]

This relaxed attitude toward personal possessions was typical throughout New England. Chrétien Le Clercq described it among the Micmac of Nova Scotia by saying that they were "so generous and liberal towards one another that they seem not to have any attachment to the little they possess, for they deprive themselves thereof very willingly and in very good spirit the very moment when they know that their friends have need of it." Europeans often interpreted such actions by emphasizing the supposed generosity of the noble savage, but the Indians' relative indifference to property accumulation is better understood as a corollary of the rest of their political and economic life. Personal goods could be easily replaced, and their accumulation made little sense for the ecological reasons of mobility we have already examined; in addition, gift giving was a crucial lubricant in sustaining power relationships within the community. As Pierre Biard noted, guests thanked their hosts by giving gifts that were expressions of relative social status, and did so "with the expectation that the host will reciprocate, when the guest comes to depart, if the guest is a Sagamore, otherwise not." Willingness to give property away with alacrity was by no means a sign that property did not exist; rather, it was a crucial means for establishing and reproducing one's position in society.[13]

When it came to land, however, there was less reason for gift giving or exchange. Southern New England Indian families enjoyed exclusive use of their planting fields and of the land on which their wigwams stood, and so might be said to have "owned" them. But neither of these were permanent possessions. Wigwams were moved every few months, and planting fields were abandoned after a number of years. Once abandoned, a field returned to brush until it was recleared by someone else, and no effort was made to set permanent boundaries around it that would hold it indefinitely for a single person. What families possessed in their fields was the *use* of them, the crops that were produced by a woman's labor upon them. When lands were traded or sold in the way Williams described, what were exchanged were usufruct rights, acknowledgments by one group that another might use an area for planting or hunting or gathering. Such rights were limited to the period of use, and they did not include many of the privileges Europeans commonly associated with ownership: a user could not (and saw no need to) prevent other village members from trespassing or gathering nonagricultural food on such lands, and had no conception of deriving rent from them. Planting fields were "possessed" by an Indian family only to the extent that it would return to them the following year. In this, they were not radically different in kind from other village lands; it was *European* rather than Indian definitions of land tenure that led the English to recognize agricultural land as the only legitimate Indian property. The Massachusetts Court made its ownership theories quite clear when it declared that "what landes any of the Indians, within this jurisdiction, have by possession or improvement, by subdueing of the same, they have just right thereunto, accordinge to that Gen: 1: 28, chap: 9: 1, Psa: 115, 16."[14]

The implication was that Indians did *not* own any other kind of land: clam banks, fishing ponds, berry-picking areas, hunting lands, the great bulk of a village's territory. (Since the nonagricultural Indians of the north had *only* these kinds of land, English theories assigned them no property rights at all.) Confusion was easy on this point, not only because of English ideologies, but because the Indians themselves had very flexible definitions of land tenure for such areas. Here again, the concept of usufruct right was crucial, since different groups of people could have different claims on the same tract of land depending on how they used it. Any village member, for instance, had the right to collect edible wild plants, cut birchbark or chestnut for canoes, or gather sedges for mats, wherever these things could be found. No special private right inhered in them. Since village lands were usually organized along a single watershed, the same was true of rivers and the coast: fish and shellfish could generally be taken anywhere, although the nets, harpoons, weirs, and tackle used to catch them—and hence sometimes the right to use the sites where these things were installed—might be owned by an individual or a kin group. Indeed, in the case of extraordinarily plentiful fishing sites—especially major inland waterfalls during the spawning runs—several villages might gather at a single spot to share the wealth. All of them acknowledged a mutual right to use the site for that specific purpose, even though it might otherwise lie within a single village's territory. Property rights, in other words, shifted with ecological use.[15]

Hunting grounds are the most interesting case of this shifting, nonagricultural land tenure. The ecological habits of different animals were so various that their hunting required a wide range of techniques, and rights to land use had to differ accordingly. The migratory birds in the ponds and salt marshes, for example, were so abundant that they could be treated much like fish: whoever killed them owned them, and hunters could range over any tract of land to do so, much like the birds themselves. (In this, Indian practices bore some resemblance to European customs governing the right of hunters, when in pursuit of game, to cross boundaries which were otherwise legally protected.) Likewise, flocks of turkeys and the deer herds were so abundant in the fall that they were most efficiently hunted by collective drives involving anywhere from twenty to three hundred men. In such cases, the entire village territory was the logical hunting region, to which all those involved in the hunt had an equal right.[16]

The same was not true, on the other hand, of hunting that involved the setting of snares or traps. The animals prey to such techniques were either less numerous, as in the case of winter deer or moose, or sedentary creatures, like the beaver, which lived in fixed locales. These were best hunted by spreading the village population over as broad a territory as possible, and so usufruct rights had to be designed to hold the overlap of trapped areas to a reasonable minimum. Roger Williams described how, after the harvest, ten or twenty men would go with their wives and children to hunting camps which were presumably organized by kin lineage groups. There, he said, "each man takes his bounds of two, three, or foure miles, where hee sets thirty, forty, or fiftie Traps, and baits his Traps with that food the Deere loves, and once in two dayes he walks his round to view his Traps."[17]

At least for the duration of the winter hunt, the kin group inhabiting a camp probably had a clear if informal usufruct right to the animals caught in its immediate area. Certainly a man (or, in the north, his wife) owned the animals captured in the traps he set, though he might have obligations to share which created *de facto* limits to his claims on them. The collective activities of a camp thus tended to establish a set of rights which at least temporarily divided the village territory into hunting areas. The problem is to know how such rights were allocated, how permanent and exclusive they were, and—most crucially—how much their interaction with the European fur trade altered them. The full discussion of this issue, which anthropologists have debated for decades, must wait for the next chapter. For now, we can conclude that, however exclusive hunting territories originally were and however much the fur trade changed them, they represented a different kind of land use—and so probably a different set of usufruct rights—than planting fields, gathering areas, or fishing sites.[18]

What the Indians owned—or, more precisely, what their villages gave them claim to—was not the land but the things that were on the land during the various seasons of the year. It was a conception of property shared by many of the hunter-gatherer and agricultural peoples of the world, but radically different from that of the invading Europeans. In nothing is this more clear than in the names they attached to their landscape, the great bulk of which related not to possession but to use. In southern New England, some of these names were agricultural. Pokanoket, in Plymouth County, Massachusetts, was "at or near the cleared lands." Anitaash Pond, near New London, Connecticut, meant, literally, "rotten corn," referring to a swampy location where corn could be buried until it blackened to create a favorite Indian delicacy. Mittineag, in Hampden County, Massachusetts, meant "abandoned fields," probably a place where the soil had lost its fertility and a village had moved its summer encampment elsewhere.[19]

Far more abundant than agricultural place-names, however, throughout New England, were names telling where plants could be gathered, shellfish collected, mammals hunted, and fish caught. Abessah, in Bar Harbor, Maine, was the "clam bake place." Wabaquasset, in Providence, Rhode Island, was where Indian women could find "flags or rushes for making mats." Azoiquoneset, also in the Narragansett Bay area, was the "small island where we get pitch," used to make torches for hunting sturgeon at night. The purpose of such names was to turn the landscape into a map which, if studied carefully, literally gave a village's inhabitants the information they needed to sustain themselves. Place-names were used to keep track of beaver dams, the rapids in rivers, oyster banks, egg-gathering spots, cranberry bogs, canoe-repairing places, and so on. Some were explicitly seasonal in their references, just as the Indian use of them was. Seconchqut Village in Dukes County, Massachusetts, was "the late spring or summer place." The Eackhonk River in Rhode Island was named to mark "the end of the fishing place," meaning the inland limit of the spring spawning runs. Unlike the

English, who most frequently created arbitrary place-names which either recalled localities in their homeland or gave a place the name of its owner, the Indians used ecological labels to describe how the land could be used.[20]

This is not to say that Indian place-names never made reference to possession or ownership. A variety of sites refer to "the boundary or ending place" which divided the territories of two different Indian villages or groups. One of the more graphic of these was Chabanakongkomuk, in Worcester, Massachusetts, a "boundary fishing place" whose name could be rendered, "You fish on your side, I fish on my side, nobody fish in the middle—no trouble." Such regions between two territories were often sites of trade: thus, Angualsicook meant the "place of barter." Most importantly, they were eventually places marking a boundary with the truly different people from across the sea. The Awannoa Path in Middlesex County, Connecticut, carried the very suggestive label "Who are you?" as a reference to "Englishmen" or "strangers."[21]

Boundaries between the Indians and these intruding "strangers" differed in fundamental ways from the ones between Indian villages, largely because the two interpreted those boundaries using very different cultural concepts. The difference is best seen in early deeds between the two groups. On July 15, 1636, the fur trader William Pynchon purchased from the Agawam village in central Massachusetts a tract of land extending four or five miles along the Connecticut River in the vicinity of present-day Springfield, leaving one of the earliest Indian deeds in American history to record the transaction. Several things are striking about the document. No fewer than thirteen Indians signed it, two of whom, Commucke and Matanchon, were evidently sachems able to act "for and in the name of al the other Indians" in the village. In defining their claims to the land being sold, they said that they acted "in the name of Cuttonus the right owner of Agaam and Quana, and in the Name of his mother Kewenusk the Tamasham or wife of Wenawis, and Niarum the wife of Coa," suggesting that both men and women had rights to the land being transferred. On the Indian side, then, an entire kin group had to concur in an action which thus probably had more to do with sovereignty than ownership.[22]

Moreover, village members evidently conceived of that action in strictly limited terms. Though they gave permission to Pynchon and his associates "for ever to trucke and sel al that ground," they made a number of revealing reservations: in addition to the eighteen coats, eighteen hatchets, eighteen hoes, and eighteen knives they received as payment, they extracted the concessions that

> they shal have and enjoy all that cottinackeesh [planted ground], or ground that is now planted; And have liberty to take Fish and Deer, ground nuts, walnuts akornes and sasachiminesh or a kind of pease.

Understood in terms of the usufruct rights discussed above, it is clear that the Indians conceived of this sale as applying only to very specific uses of the land. They gave up none of their most important hunting and gathering privileges, they retained right to their cornfields, and evidently intended to keep living on the land much as they had done before. The rights they gave Pynchon were apparently to occupy the land jointly with them, to establish a village like their own where cornfields could be planted, to conduct trade there, and perhaps to act as a superior sachem who could negotiate with other villages about the land so long as he continued to recognize the reserved rights of the Agawam village. The Agawam villagers gave up none of their sovereignty over themselves, and relinquished few of their activities on the land. What they conferred on Pynchon was a right of ownership identical to their own: not to possess the land as a tradeable commodity, but to use it as an ecological cornucopia. Save for cornfields, no Indian usufruct rights were inherently exclusive, and transactions such as this one had more to do with sharing possession than alienating it.[23]

On the English side, the right "for ever to trucke and sel al that ground" of course carried rather different connotations. In the first place, the transaction was conducted not by a sovereign kin group but by a trading partnership operating under the much larger sovereignty of the Massachusetts Bay Company and the English Crown. None of the three partners who acquired rights to the land—William Pynchon, Henry Smith, or Jehu Burr—was actually present at the transaction, which was conducted for them by several men in their employ. Insofar as we can make a

valid distinction, what the Indians perceived as a political negoti-
ation between two sovereign groups the English perceived as an
economic transaction wholly within an English jurisdiction. As
we have seen, Massachusetts recognized that Indians might have
limited natural rights to land, and so provided that such rights
could be alienated *under the sanctions of Massachusetts law.* No question of an Indian village's own sanctions could arise, for the
simple reason that Indian sovereignty was not recognized. The
Massachusetts Bay Company was careful very early to instruct
its agents on this point, telling them "to make composition with
such of the salvages as did pretend any tytle or lay clayme to any
of the land." Indian rights were not real, but pretended, because
the land had already been granted the company by the English
Crown.[24]

Land purchases like Pynchon's were thus interpreted under
English law, and so were understood as a fuller transfer of rights
than Indian communities probably ever intended. Certainly Pynchon's deed is unusual in even mentioning rights reserved to the
Indians. Later deeds describe exchanges in which English purchasers appeared to obtain complete and final ownership rights,
however the Indian sellers may have understood those exchanges. In 1637, for instance, John Winthrop received lands in
Ipswich, Massachusetts, from the Indian Maskonomett, who declared that "I doe fully resigne up all my right of the whole towne
of Ipswich as farre as the bounds thereof shall goe all the woods
meadowes, pastures and broken up grounds unto the said John
Winthrop in the name of the rest of the English there planted."
Deeds in eastern Massachusetts—when they existed at all—typically took this form, extinguishing all Indian rights and transferring them either to an English purchaser or, as in this case, to an
English group with some corporate identity. As the English understood these transactions, what was sold was not a bundle of
usufruct rights, applying to a range of different "territories," but
the land itself, an abstract area whose bounds in theory remained
fixed no matter what the use to which it was put. Once the land
was bounded in this new way, a host of ecological changes followed almost inevitably.[25]

European property systems were much like Indian ones in
expressing the ecological purposes to which a people intended to
put their land; it is crucial that they not be oversimplified if their

contribution to ecological history is to be understood. The popular idea that Europeans had private property, while the Indians did not, distorts European notions of property as much as it does Indian ones. The colonists' property systems, like those of the Indians, involved important distinctions between sovereignty and ownership, between possession by communities and possession by individuals. They too dealt in bundles of culturally defined rights that determined what could and could not be done with land and personal property. Even the fixity they assigned to property boundaries, the quality which most distinguished them from Indian land systems, was at first fuzzier and less final than one might expect. They varied considerably depending on the region of England from which a group of colonists came, so that every New England town, like every Indian village, had idiosyncratic property customs of its own. All of these elements combined to form what is usually called "the New England land system." The phrase is misleading, since the "system" resided primarily at the town level and was in fact many systems, but there were nevertheless common features which together are central to the subject of this book. Their development was as much a product as a cause of ecological change in colonial New England.

Colonial claims to ownership of land in New England had two potential sources: purchases from Indians or grants from the English Crown. The latter tended quickly to absorb the former. The Crown derived its own claim to the region from several sources: Cabot's "discovery" of New England in 1497-98; the failure of Indians adequately to subdue the soil as Genesis 1.28 required; and from the King's status—initially a decidedly speculative one—as the first Christian monarch to establish colonies there. Whether or not a colony sought to purchase land from the Indians—something which Plymouth, Connecticut, and Rhode Island, in the absence of royal charters, felt compelled as a matter of expediency or ethics to do—all New England colonies ultimately derived their political rights of sovereignty from the Crown.[26]

The distinction between sovereignty and ownership is crucial here. When a colony purchased land from Indians, it did so under its own system of sovereignty: whenever ownership rights were deeded and purchased, they were immediately incorporated into

English rather than Indian law. Indian land sales, operating as they did at the interface of two different sovereignties, one of which had trouble recognizing that the other existed, thus had a potentially paradoxical quality. Because Indians, at least in the beginning, thought they were selling one thing and the English thought they were buying another, it was possible for an Indian village to convey what it regarded as identical and nonexclusive usufruct rights to several different English purchasers. Alternatively, several different Indian groups might sell to English ones rights to the same tract of land. Uniqueness of title as the English understood it became impossible under such circumstances, so colonies very early tried to regulate the purchase of Indian lands. Within four years of the founding of Massachusetts Bay, the General Court had ordered that "noe person whatsoever shall buy any land of any Indean without leave from the Court." The other colonies soon followed suit. The effect was not only to restrict the right of English individuals to engage in Indian land transactions but—more importantly, given the problem of sovereignty—to limit the rights of Indians to do so as well. Illegal individual sales nevertheless persisted, and titles in some areas became so confused that the Connecticut Court in 1717 made a formal declaration:

> That all lands in this government are holden of the King of Great Britain as the lord of the fee: and that no title to any lands in this Colony can accrue by any purchase made of Indians on pretence of their being native proprietors thereof.

Even by the late seventeenth century, Indian lands were regarded as being entirely within English colonial jurisdiction; indeed, the logic of the situation seemed to indicate that, for Indians to own land at all, it had first to be granted them by the English Crown.[27]

If all colonial lands derived from the Crown, how did this affect the way they were owned and used? As with an Indian sachem, albeit on a larger and more absolute scale, the King did not merely possess land in his own right but also represented in his person the collective sovereignty which defined the system of property rights that operated on that land. In the case of the

Massachusetts Bay Company's charter, the King conferred the lands of the grant "as of our manor of Eastgreenewich, in the County of Kent, in free and common Socage, and not in Capite, nor by knightes service." Land tenure as of the manor of East Greenwich put a colony under Kentish legal custom and was the most generous of feudal grants, involving the fewest obligations in relation to the Crown. It was ideally suited to mercantile trading companies, since it allowed easy alienation of the land and did not impose the burden of feudal quitrents on its holders. Both of these features made Kentish tenure attractive to would-be settlers and promoted the early development of a commercial market in land. As opposed to tenure in capite or by knight's service, which carried various civil and military obligations for their holders, free and common socage—in some senses, the least feudal of medieval tenures—conceived of land simply as property carrying an economic rent, a rent which was often negligible. In Massachusetts, the Crown's only claim was to receive one-fifth of all the gold and silver found there. Given New England geology, the burden did not prove onerous.[28]

The royal charter drew a set of boundaries on the New England landscape. Unlike those of the Indians, these were not "boundary or ending places" between the territories of two peoples. Rather, they were defined by lines of latitude—40 and 48 degrees north—that in theory stretched from "sea to sea." Between those lines, the Massachusetts Bay Company was given the right

> TO HAVE and to houlde, possesse, and enjoy all and singuler the aforesaid continent, landes, territories, islands, hereditaments, and precincts, seas, waters, fishings, with all and all manner their commodities, royalties, liberties, prehemynences, and profitts that should from thenceforth arise from thence, with all and singuler their appurtenances, and every parte and parcell thereof, unto the saide Councell and their successors and assignes for ever.

It was an enormous grant, no doubt in part because the King's personal claim to the territory was so tenuous. For our purposes, its significance lies in the sweeping extent and abstraction of its rights and boundaries, its lack of concern for the claims of existing inhabitants, its emphasis on the land's profits and commodities, and its intention that the land being granted could and would remain so bounded "forever." In all of these ways, it implied conceptions of land tenure drastically different from those of the Indians.[29]

Because the King's grant was so permissive, and gave so little indication as to how land should be allocated within the new colony, the company and its settlers found themselves faced with having to devise their own method for distributing lands. Initially, the company thought to make grants to each shareholder and settler individually, as had been done in Virginia, but this idea was rapidly—though not completely—replaced with grants to groups of settlers acting together as towns. The founding proprietors of each town were collectively granted an average of about six square miles of land, and from then on were more or less free to dispose of that land as they saw fit. In terms of sovereignty, the chief difference between Indian and English villages lay in the formal hierarchy by which the latter derived and maintained their sovereign rights. But in terms of ownership —the way property and usufruct rights were distributed *within* a village—the two differed principally in the ways they intended ecologically to use the land. When the Agawam villagers reserved hunting and gathering rights in their deed to William Pynchon, they revealed how they themselves thought that particular tract of land best used. Likewise, John Winthrop's deed to Ipswich— clearly an English rather than an Indian document—in speaking of "woods meadowes, pastures and broken up grounds," betrayed the habits of thought of an English agriculturalist who was accustomed to raising crops, building fences, and keeping cattle. Conceptions of land tenure mimicked systems of ecological use.[30]

The proprietors of a new town initially held all land in common. Their first act was to determine what different types of land were present in their territory, types which were understood to be necessary to English farming in terms of the categories mentioned in Winthrop's deed: forested lands for timber and firewood, grassy areas for grazing, salt marshes for cutting hay, potential planting fields, and so on. Like their Indian counterparts, English villages made their first division of land to locate where houses and cornfields should be; unlike the Indians, that division was conducted formally and was intended to be a permanent one, the land passing forever into private hands. Land was allocated to inhabitants using the same biblical philosophy that had justified taking it from the Indians in the first place: individuals should only possess as much land as they were able to subdue and make productive. The anonymous "Essay on the Ordering of Towns" declared that each inhabitant be given "his due proportion, more or lesse according unto his present or apparent future occasion of Imployment." A person with many servants and cattle could "improve" more land than one who had few, and so was granted more land, although the quantities varied from town to town. In this way, the social hierarchy of the English class system was reproduced, albeit in modified form, in the New World. Grants of house lots and planting grounds were followed by grants of pastures, hay meadows, and woodlots, all allocated on the same basis of one's ability to use them.[31]

In these and later grants as well, the passage of land from town commons to individual property was intended to create permanent private rights to it. These rights were never absolute, since both town and colony retained sovereignty and could impose a variety of restrictions on how land might be used. Burning might be prohibited on it during certain seasons of the year. A grant might be contingent on the land being used for a specific purpose —such as the building of a mill—and there was initially a requirement in Massachusetts that all land be improved within three years or its owner would forfeit rights to it. Regulations might forbid land from being sold without the town's permission. But, compared with Indian villages, grants made by New England towns contemplated much more extensive privileges for each individual landholder, with greater protection from trespass and more exclusive rights of use. The "Essay on the Ordering of Towns" saw such private ownership as the best way to promote fullest use of the land: "he that knoweth the benefit of incloseing," it said, "will omit noe dilligence to brenge him selfe into an inclusive condicion, well understanding that one acre inclosed, is much more beneficiall than 5 falling to his share in Common."[32]

Different towns acted differently at first in relation to their common lands, their behavior usually depending on the land practices of the regions of England from which their inhabitants came. Some settlers, like those of Rowley or Sudbury, came from areas with open-field systems, where strong manorial control had been exercised over lands held in common by peasant farmers. They initially re-created such systems in New England, making relatively few small divisions of common holdings, regulating closely who could graze and gather wood on unenclosed land, and not engaging extensively in the buying or selling of real estate. Settlers in towns like Ipswich or Scituate, on the other hand, came from English regions where closed-field systems gave peasant proprietors more experience with owning their lands in severalty. They proved from the start to be much interested in transferring lands from common to private property as rapidly as possible, so that their land divisions were more frequent and involved more land at an earlier date. In these towns, a market in real estate developed very early, both to allow the consolidation of scattered holdings and to facilitate limited speculative profits in land dealings.[33]

In the long run, it was this latter conception of land—as private commodity rather than public commons—that came to typify New England towns. Initial divisions of town lands, with their functional classifications of woodlot and meadow and cornfield, bore a superficial resemblance to Indian usufruct rights, since they seemed to define land in terms of how it was to be used. Once transferred into private hands, however, most such lands became abstract parcels whose legal definition bore no inherent relation to their use: a person owned everything on them, not just specific activities which could be conducted within their boundaries. Whereas the earliest deeds tended to describe land in terms of its topography and use—for instance, as the mowing field between a certain two creeks—later deeds described land in terms of lots held by adjacent owners, and marked territories using the surveyor's abstractions of points of the compass and metes and bounds. Recording systems, astonishingly sloppy in the beginning because there was little English precedent for them, became increasingly formalized so that boundaries could be more precisely defined. Even Indian deeds showed this transformation. The land Pynchon purchased from the Agawam village was vaguely defined in terms of cornfields, meadows, and the Connecticut River; an eighteenth-century deed from the same county, on the other hand, transferred rights to two entire townships which it defined precisely but abstractly as "the full Contents of Six miles in Weadth and Seven miles in length," starting from a specified point.[34]

The uses to which land could be put vanished from such descriptions, and later land divisions increasingly ignored actual topography. What was on the land became largely irrelevant to its legal identity, even though its contents—and the rights to them—might still have great bearing on the price it would bring if sold. Describing land as a fixed parcel with purely arbitrary boundaries made buying and selling it increasingly easy, as did the recording systems—an American innovation—which kept track of such transactions. Indeed, legal descriptions, however abstracted, had little effect on everyday life *until* land was sold. People did not cease to be intimately a part of the land's ecology simply by reason of the language with which their deeds were written. But when it came time to transfer property rights, those deeds allowed the alienation of land as a commodity, an action with important ecological consequences. To the abstraction of legal boundaries was added the abstraction of price, a measurement of property's value assessed on a unitary scale. More than anything else, it was the treatment of land and property as commodities traded at market that distinguished English conceptions of ownership from Indian ones.

To present these arguments in so brief a compass is of course to oversimplify. Western notions of property, commodity, and market underwent a complex development in both Europe and America over the course of the seventeenth and eighteenth centuries, one which did not affect all people or places in the same way or at the same time. Peasant land practices which had their origins in the manorial customs of feudal England were not instantly transformed into full-fledged systems of production for market simply by being transferred to America. Many communities produced only a small margin of surplus beyond their own needs, and historians have often described them as practicing "subsistence agriculture" for this reason. When seventeenth-century New England towns are compared with those of the nineteenth century, with their commercial agriculture, wage workers, and urban industrialism, the transition between the two may well seem to be that from a subsistence to a capitalist society. Certainly Marxists wedded to a definition of capitalism in terms of relations between labor and capital must have trouble seeing it in the first New England towns. Most early farmers owned their own land, hired few wage laborers, and produced mainly for their own use. Markets were hemmed in by municipal regulations, high transportation costs, and medieval notions of the just price. In none of these ways does it seem reasonable to describe colonial New England as "capitalist."[35]

And yet when colonial towns are compared not with their industrial successors but with their Indian predecessors, they begin to look more like market societies, the seeds of whose capitalist future were already present. The earliest explorers' descriptions of the New England coast had been framed from the start in terms of the land's commodities. Although an earlier English meaning of the word "commodity" had referred simply to articles which were "commodious" and hence useful to people —a definition Indians would readily have understood—that meaning was already becoming archaic by the seventeenth century. In its place was the commodity as an object of commerce, one by definition owned for the sole purpose of being traded away at a profit. ("Profit" was another word that underwent a comparable evolution at about the same time: to its original meaning of the benefits one derived from using a thing was added the gain one made by selling it.) Certain items of the New England landscape—fish, furs, timber, and a few others—were thus selected at once for early entrance into the commercial economy of the North Atlantic. They became valued not for the immediate utility they brought their possessors but for the price they would bring when exchanged at market. In trying to explain ecological changes related to these commodities, we can safely point to market demand as the key causal agent.[36]

The trade in commodities involved only a small group of merchants, but they exercised an influence over the New England economy beyond their numbers. Located principally in the coastal cities, they rapidly came to control shipping and so acted as New England's main link to the Atlantic economy. Because of their small numbers, it might reasonably be argued that the market sector of the New England economy was a tiny isolated segment relatively unconnected to the subsistence production of peasant communities in the towns. Certainly we should make a distinction between ecological changes resulting directly from the activities of merchants and those caused by the less market-oriented activities of farmers. But the farmers had their own involvement in the Atlantic economy, however distant it might have been. Even if they produced only a small surplus for market, they nevertheless used it to buy certain goods from the merchants—manufactured textiles, tropical foodstuffs, guns, metal tools—which were essential elements in their lives. The grain and meat which farmers sold, if not shipped to Caribbean and European markets, were used to supply port cities and the "invisible trade" of colonial shipping. Not all of this commodity movement was voluntary. Town and colony alike assessed farmers for their landholdings and so siphoned off taxes which were used to run government and conduct trade. Although taxes bore some resemblance to political tributes in Indian societies, the latter were not based on possession of land and did not reinforce the sense that land had an intrinsic money value. Taxes thus had the important effect of forcing a certain degree of colonial production beyond the level of mere "subsistence," and orienting that surplus toward market exchange.[37]

But the most important sense in which it is wrong to describe colonial towns as subsistence communities follows from their inhabitants' belief in "improvement," the concept which was so crucial in their critique of Indian life. The imperative here was not just the biblical injunction to "fill the earth and subdue it." Colonists were moved to transform the soil by a property system that taught them to treat land as capital. Fixed boundaries and the liberties of "free and common socage" assured a family that improvements belonged to them and to their heirs. The existence of commerce, however marginal, led them to see certain things on the land as merchantable commodities. The visible increase in livestock and crops thus translated into an abstract money value that was reflected in tax assessments, in the inventories of estates, and in the growing land market. Even if a colonist never sold an improved piece of property, the increase in its hypothetical value at market was an important aspect of the accumulation of wealth. These tendencies were apparent as early as the 1630s. When English critics claimed that colonists had lost money by moving their wealth to New England, the colonists replied that they had simply transformed that money into physical assets. The author of *New England's First Fruits* declared that the colonists' "estates

now lie in houses, lands, horses, cattel, corne, etc. though they
have not so much money as they had here [in England], and so
cannot make appearance of their wealth to those in *England,* yet
they have it still, so that their estates are not lost, but changed."
Here was a definition of transformable wealth few precolonial
Indians would probably have recognized: if labor was not yet an
alienated commodity available for increasing capital, land was.
"The staple of America at present," wrote the British traveler
Thomas Cooper in the late eighteenth century, "consists of Land,
and the immediate products of land."[38]

Perhaps the best single summary of this view is John Locke's
famous chapter on property in the *Two Treatises of Government.*
Locke sought to explain how people came to possess unequal
rights to a natural abundance he supposed had originally been
held in common; to accomplish this task, he explicitly contrasted
the societies of Europe with those of the American Indians. "In
the beginning," he said, "all the world was America." In that
original state, possession was directly related to the labor one
spent in hunting and gathering: one could own whatever one
could use before it spoiled. What enabled people to accumulate
wealth beyond the limits of natural spoilage was something
Locke called "money." Bullionist that he was, he thought of
money as gold and silver which could be stored as a source of
permanent value without fear of spoiling. But the way he actually used the word, "money" was an odd hybrid between a simple
medium of exchange that measured the value of commodities,
and *capital,* the surplus whose accumulation was the motor of
economic growth. It was capital—the ability to store wealth in
the expectation that one could increase its quantity—that set
European societies apart from precolonial Indian ones. As Locke
said:

> Where there is not something both lasting and scarce, and
> so valuable to be hoarded up, there Men will not be apt to
> enlarge their *Possessions of Land,* were it never so rich, never
> so free for them to take. For I ask, What would a Man value
> Ten Thousand or a Hundred Thousand Acres of excellent
> *Land,* ready cultivated, and well stocked too with Cattle, in
> the middle of the in-land Parts of *America,* where he had no
> hopes of Commerce with other Parts of the World, to draw

Money to him by the Sale of the Product? It would not be worth the inclosing, and we should see him give up again to the wild Common of Nature, whatever was more than would supply the Conveniences of Life to be had there for him and his Family.

New England had not returned to the "wild Common of Nature" but had in fact abandoned it. However incomplete Locke's analysis of why that had happened, and however inaccurate his anthropological description of Indian society, his emphasis on the market was sound. It was the attachment of property in land to a marketplace, and the accumulation of its value in a society with institutionalized ways of recognizing abstract wealth (here we need not follow Locke's emphasis on gold and silver), that committed the English in New England to an expanding economy that was ecologically transformative.[39]

Locke carries us full circle back to Thomas Morton's riddle. His characterization of the Indians as being "rich in Land, and poor in all the Comforts of Life," bore a close resemblance to the comparisons of Indians with English beggars which Morton had sought to refute. Locke posed the riddle of Indian poverty as clearly as anyone in the seventeenth century. He described them as a people

whom Nature having furnished as liberally as any other people, with the materials of Plenty, *i.e.* a fruitful Soil, apt to produce in abundance, what might serve for food, rayment, and delight; yet for want of improving it by labour, have not one hundredth part of the Conveniences we enjoy: And a King of a large fruitful Territory there feeds, lodges, and is clad worse than a day Labourer in *England.*

Because the Indians lacked the incentives of money and commerce, Locke thought, they failed to improve their land and so remained a people devoid of wealth and comfort.[40]

What Locke failed to notice was that the Indians did not recognize themselves as poor. The endless accumulation of capital which he saw as a natural consequence of the human love for wealth made little sense to them. Marshall Sahlins has pointed out that there are in fact two ways to be rich, one of which was rarely recognized by Europeans in the seventeenth century. "Wants," Sahlins says, "may be 'easily satisfied' either by producing much or desiring little." Thomas Morton was almost alone among his contemporaries in realizing that the New England Indians had chosen this second path. As he said, on their own understanding, they "lived richly," and had little in the way of either wants or complaints. Pierre Biard, who also noticed this fact about the Indians, extended it into a critique of *European* ways of life. Indians, he said, went about their daily tasks with great leisure,

> for their days are all nothing but pastime. They are never in a hurry. Quite different from us, who can never do anything without hurry and worry; worry, I say, because our desire tyrannizes over us and banishes peace from our actions.

Historians often read statements like this as myths of the noble savage, and certainly they are attached to that complex of ideas in European thought. But that need not deny their accuracy as descriptions of Indian life. If the Indians considered themselves happy with the fruits of relatively little labor, they were like many peoples of the world as described by modern anthropologists.[41]

Thomas Morton had posed his riddle knowing full well that his readers would recognize its corollary: if Indians lived richly by wanting little, then might it not be possible that Europeans lived poorly by wanting much? The difference between Indians and Europeans was not that one had property and the other had none; rather, it was that they loved property differently. Timothy Dwight, writing at the beginning of the nineteenth century, lamented the fact that Indians had not yet learned the love of property. "Wherever this can be established," he said, "Indians may be civilized; wherever it cannot, they will still remain Indians." The statement was truer than he probably realized. Speaking strictly in terms of precolonial New England, Indian conceptions of property were central to Indian uses of the land, and Indians could not live as Indians had lived unless the land was owned as Indians had owned it. Conversely, the land could not long remain unchanged if it were owned in a different way. The sweeping alterations of the colonial landscape which Dwight himself so shrewdly described were testimony that a people who loved property little had been overwhelmed by a people who loved it much.[42]

5

COMMODITIES OF
THE HUNT

The tension between Indian and European property systems did
not become instantly apparent the moment the first European
visited New England. Indeed, for well over a century before
English settlement began in Massachusetts, Europeans and Indians engaged in a largely unrecorded trade which suggested more
possibilities for cooperation than for conflict between their respective economies. Hunters and sailors encountering one another on the coasts of Maine, Nova Scotia, and the St. Lawrence
discovered very early that they had valuable things to trade:
metal goods, weaponry, articles of clothing, and ornamental objects on the part of the Europeans; furs and skins on the part of
the Indians. For the Europeans, such trade began as a casual
adjunct to the cod fisheries, but in the second half of the sixteenth
century, with the rising popularity of felt hats and the decline in
European fur production, North American furs became a principal object of trade in their own right. For the Indians, that trade
marked a new involvement in an alien commercial economy, as
well as the onset of complicated shifts in their ecological circumstances.[1]

It is important to underscore how little we know of this early fur trade and its effects: the various Indian peoples of New England undoubtedly started interacting with European visitors neither at the same time nor in the same way. As early as 1524, Verrazzano found Indians along the Maine coast who were only too familiar with the ways of European sailors. Although they actively sought to trade furs for knives, hooks, and other metal goods, they were nevertheless careful to direct Verrazzano's men to meet them at the rockiest and most dangerous part of the shore, where landing was impossible. There, as Verrazzano reported, "they sent us what they wanted to give on a rope, continually shouting to us not to approach the land," and communicating great hostility. This was in marked contrast to the inhabitants of Narragansett Bay, who apparently had had rather less experience with Europeans. They not only welcomed Verrazzano wholeheartedly but seemed indifferent to his offers of trade. "They did not appreciate cloth of silk and gold," Verrazzano wrote, "nor even of any other kind, nor did they care to have them; the same was true for metals like steel and iron, for many times when we showed them some of our arms, they did not admire them, nor ask for them, but merely examined their workmanship." Northern Indians, who lived closer to the fishing banks, learned of European trade goods and military methods before their southern counterparts; likewise, inhabitants of the coast began European contacts before those of interior villages.[2]

Nevertheless, by the beginning of the seventeenth century, every recorded European exploration found Indians in villages all along the New England coast eager for trade. On the island of Cuttyhunk in 1602, Bartholomew Gosnold obtained the skins of beavers, otters, martens, foxes, rabbits, seals, and deer in exchange for knives and what he called "trifles." In 1605, Champlain was greeted on the Penobscot River by thirty Indians led by a sachem named Bashaba, who assured him that "no greater good could come to them than to have our friendship." Champlain was told that "they desired to live in peace with their enemies, and that we should dwell in their land, in order that they might in future more than ever before engage in hunting beavers, and give us a part of them in return for our providing them with things which they wanted." Already in 1605, Bashaba's speech displayed the extent to which Indians were orienting their economic activity to enable them to trade with Europeans. Bashaba made clear that he understood the demand for beaver on European markets, that he and his followers were prepared to increase their production to meet that demand, and that he saw friendship with Europeans as a way both of obtaining trade goods and of shifting military balances among Indian villages.[3]

These were important lessons, to which we must soon return. For the moment, however, the most significant thing about them was that they were learned primarily not from men like Champlain and Gosnold but from dozens of unknown visitors who left no record of their trips. We know next to nothing about most of the Europeans who journeyed to New England in the sixteenth and early seventeenth centuries, and we can make only the crudest of inferences about how Indians responded to them. Yet there can be no doubt that contacts between the two groups were extensive. Explorers who were greeted by Indians speaking French, English, or Basque could have few illusions about being the first European visitors to an area. When the Pilgrims first landed on Cape Cod in 1620, they discovered "a place like a grave" covered with wooden boards. Digging it up, they found layer upon layer of household goods, the personal possessions that Indians ordinarily buried with their dead: mats, bowls, trays, dishes, a bow, and two bundles. In the smaller of the two bundles was a quantity of sweet-smelling red powder in which were the bones of a young child, wrapped in beads and accompanied by an undersized bow. Still, what troubled the graverobbers were not these Indian things, many of which they took, but the contents of the larger bundle. There, in the same red powder, were the remnants of a man: some of the flesh remained on the bones, and they realized with a shock that "the skull had fine yellow haire still on it." With the bones, "bound up in a saylers canvas Casacke, and a payre of cloth breeches," were a knife, a needle, and "two or three old iron things," evidently the dead man's most personal belongings. A blond European sailor, shipwrecked or abandoned on the Massachusetts coast, had lived as an Indian, had perhaps fathered an Indian child, and had been buried in an Indian grave. His circumstances may or may not have been unusual—even this we cannot know—but they betokened an already long and continuing exchange between peoples on opposite sides of the Atlantic.[4]

It was anonymous Europeans like the dead sailor in the Cape Cod grave who helped bring about the single most dramatic ecological change in Indian lives, one whose full significance historians have only recently come to understand. Of all the many organisms Europeans carried to America, none of them were more devastating to the Indians than the Old World diseases, what W. H. McNeill has called the microparasites. The ancestors of New England Indians had reached North America perhaps twenty or thirty thousand years before the Europeans, arriving via a land bridge at the Bering Straits during the most recent of the glacial epochs. In the course of their migrations, they failed to bring with them many of the illnesses that were common elsewhere in the world. Their low population densities and their having lived for extended periods under semiarctic conditions served to filter out microorganisms which required large host populations and more temperate climates to survive. Their lack of domesticated animals, especially the grazing ungulates, such as swine, cattle, and horses, which share many diseases with humans, also helped to reduce the number of pathogens they brought with them from Eurasia. As a result, the American Indians were blessed by the absence not only of diseases that Europeans ordinarily experienced in childhood, such as chicken pox and measles, but also of more lethal organisms that were epidemic in the Old World: smallpox, influenza, plague, malaria, yellow fever, tuberculosis, and several others. As William Wood said, the New England Indians were a people of "lusty and healthful bodies, not experimentally knowing the catalogue of those health-wasting diseases which are incident to other countries."[5]

It was, of course, a blessing that proved all too quickly to be a curse. For hundreds of generations, Indian babies had grown to adulthood with no experience of these illnesses, so that Indian mothers transferred to their infants none of the antibodies which might have provided some measure of immunity against them. What the Indians lacked was not so much *genetic* protection from Eurasian disease—though this may have been a partial factor—as the historical experience as a population to maintain *acquired* immunities from generation to generation: each new generation that failed to encounter a disease was left with less protection against it. As a result, European diseases struck Indian villages with horrible ferocity. Mortality rates in initial onslaughts were rarely less than 80 or 90 percent, and it was not unheard of for an entire village to be wiped out. From the moment of their first contact with an Old World pathogen, Indian populations experienced wave upon wave of epidemics as new diseases made their appearance or as new nonimmune generations came of age. A long process of depopulation set in, accompanied by massive social and economic disorganization.[6]

How early this began in New England is impossible to know. The long sea voyage across the Atlantic acted as a disease filter in its own right, since epidemic diseases had usually run their course on board ship before sailors reached North America. Many visits probably occurred with no disease organisms at all being passed on to Indians, and the earliest transfers may well have been of endemic illnesses—diarrheas, dysentery, venereal diseases, respiratory viruses, tuberculosis—which had remained nonvirulent in the shipboard population. Those Indians who saw the most of European visitors were obviously most at risk to acquire any disease that survived the sea journey, and so the inhabitants of northern fur-trading areas probably began to suffer from the new diseases first. By 1616, Pierre Biard could write of disease as being a regular visitor to the Indians of Maine and Nova Scotia. "They are astonished," he said,

> and often complain that, since the French mingle with and carry on trade with them, they are dying fast, and the population is thinning out. For they assert that, before this association and intercourse, all their countries were very populous, and they tell how one by one the different coasts, according as they have begun to traffic with us, have been more reduced by disease.

Biard's description suggests a significant increase in mortality rates among the northern Indians, but does not convey the catastrophic scale of epidemics that occurred elsewhere. The low population densities of the northern hunters probably protected them somewhat from certain diseases (cholera, for instance) that needed a minimum host population in order to reproduce themselves, and may also have limited the diseases that did occur to sporadic outbreaks. Nevertheless, northern death sentences were postponed rather than annulled, and the effects of disease were simply spread over a long period of slow attrition.[7]

Although the corn-growing Indians of southern New England were less involved in the fur trade, their much greater population densities meant that when disease was finally introduced to them, its effects were explosive. The first recorded epidemic in the south began in 1616 and raged for three years on the coast between Cape Cod and Penobscot Bay, reaching villages perhaps twenty or thirty miles inland but sparing both the deeper interior and the coast west of Narragansett Bay. Although contemporary observers described it as "the plague," New England lacked the rats and human population densities necessary to sustain that disease. Chicken pox seems a more likely cause, since its virus requires only a small host population in order to remain in circulation; it could have traveled in latent form across the Atlantic in the body of a European sailor who might then have developed shingles from which Indians could easily have become infected. Whichever disease the Indians caught, its effects are well documented. Thomas Morton told of villages in which only a single inhabitant was left alive. So many died that no one remained to bury the corpses, and crows and wolves feasted on them where they lay. When colonists arrived a few years after the epidemic had spent itself, they found such quantities of bleached bones and skulls that, as Morton said, "it seemed . . . a new found Golgotha."[8]

Southern New England Indians continued to experience serious outbreaks of disease during the 1620s, and in 1633 they were visited by smallpox, one of the most lethal of European killers. The 1633 epidemic saw mortalities in many villages reach 95 percent. This time villages in the interior and along the coast west of Narragansett Bay fell victim to the viral fury as much as those elsewhere, perhaps an indication of the extent to which their trade connections had been expanded since 1616. William Bradford, Governor of the Plymouth Colony, has left the fullest description of the horrors brought by the disease. "For want of bedding and linen and other helps," he wrote,

> they fall into a lamentable condition as they lie on their hard mats, the pox breaking and mattering and running one into another, their skin cleaving by reason thereof to the

mats they lie on. When they turn them, a whole side will flay off at once as it were, and they will be all of a gore blood, most fearful to behold. And then being very sore, what with cold and other distempers, they die like rotten sheep.[9]

There was little Indians could do to protect themselves from the epidemics. Whereas they had previously dealt with their sick companions by gathering at their bedside to sit through the illness with them, they quickly learned that the new diseases could be escaped only by casting aside family and community ties and fleeing. "So terrible is their apprehension of an infectious disease," wrote Roger Williams in the 1640s, "that not only persons, but the Houses and the whole Towne takes flight." With only the sick left to help the sick, even those who might otherwise have survived an epidemic were doomed. Bradford said that

they were in the end not able to help one another, no not to make a fire nor to fetch a little water to drink, nor any to bury the dead. But would strive as long as they could, and when they could procure no other means to make fire, they would burn the wooden trays and dishes they ate their meat in, and their very bows and arrows. And some would crawl out on all fours to get a little water, and sometimes die by the way and not be able to get in again.[10]

Social disorganization compounded the biological effects of disease. Once villages were attacked by a new pathogen, they often missed key phases in their annual subsistence cycles—the corn planting, say, or the fall hunt—and so were weakened when the next infection arrived. Worse, hungry times that had always been normal in precolonial Indian society—for instance, the late winter among northern hunters—became lethal when accompanied by the new diseases. Chronic illnesses gained their foothold in this way, and broke out whenever Indian populations became particularly susceptible to them. Tuberculosis had become common by the end of the seventeenth century, and influenza in combination with pneumonia recurred regularly in epidemic proportions. Measles, typhus, dysentery, and syphilis all became endemic and contributed to the general decline in Indian populations. As a result, in the first seventy-five years of the seventeenth century, the total number of Indians in New England fell precipitously from well over 70,000 to fewer than 12,000. In some areas, the decline was even more dramatic: New Hampshire and Vermont were virtually depopulated as the western Abenaki declined from perhaps 10,000 to fewer than 500.[11]

The epidemics disrupted most of the networks of kinship and authority that had previously organized Indian lives. When Bradford described a village in which "the chief sachem himself now died and almost all his friends and kindred," he was depicting a phenomenon that took place in many Indian communities. Villages which had lost their sachems and whose populations had declined twentyfold were often no longer viable entities; surviving Indians were forced to move to new villages and create new political alignments. Depopulation and alliances with Europeans gave ambitious individuals who had lacked high rank before an epidemic opportunities to assume new leadership roles. Squanto, for instance, later to become the Pilgrims' interpreter, was the only survivor from his village at the end of the 1616-19 epidemic. A man without a community, "whose ends," as Edward Winslow said, "were only to make himself great in the eyes of his countrymen, by means of his nearness and favor with us," he consciously sought to undermine the authority of the neighboring sachem Massasoit. One of his devices for doing this was to convince other Indians that the colonists "had the plague buried in our storehouse; which, at our pleasure, we could send forth to what place or people we would, and destroy them therewith, though we stirred not from home." This was a particularly blatant attempt to use disease to amass political power; the Indians' willingness to believe it testifies to both their fear and their well-grounded suspicion that the new illnesses were of European origin. But the mere fact of depopulation promoted conditions of turmoil which enabled new leaders to emerge in the ensuing political vacuum.[12]

The social disruption brought by the epidemics was not limited to political leaders. Indian doctors, or powwows, found their ordinary healing practices useless against so potent a biological assault. Indeed, as the Puritan historian Edward Johnson said, the "powwows themselves were oft smitten with deaths stroke." European pathogens thus served to undermine the spiritual and religious practices of Indian communities. John Winthrop, in speaking of victims of smallpox in 1633, said that "divers of them, in their sickness, confessed that the Englishmen's God was a good God; and that, if they recovered, they would serve him." Conversions of this sort were often a kind of hedging of bets with little lasting consequence, but they suggest some of the spiritual trauma brought by the enormous mortalities. As Robert Cushman said of the Indians around Plymouth, "Those that are left, have their courage much abated, and their countenance is dejected, and they seem as a people affrighted."[13]

Indian depopulation as a result of European diseases ironically made it easier for Europeans to justify taking Indian lands. If the English believed that cornfields were the only property Indians had improved sufficiently to own, the wiping out of a village—and the subsequent abandonment of its planting fields—eliminated even this modest right. Over and over again, New England towns made their first settlements on the sites of destroyed Indian villages. Plymouth, for instance, was located "where there is a great deale of Land cleared, and hath beene planted with Corne three or foure yeares agoe"—planted, in fact, just before the 1616 epidemic broke out. More than fifty of the earliest settlements had similar locations, thus saving their inhabitants much initial work in clearing trees. To Puritans, the epidemics were manifestly a sign of God's providence, "in sweeping away great multitudes of the natives . . . that he might make room for us there." John Winthrop saw this "making room" as a direct conveyance of property right: "God," he said, "hath hereby cleared our title to this place."[14]

As Indian villages vanished, the land on which they had lived began to change. Freed from the annual burnings and soon to be subject to an entirely different agricultural regime, the land's transformations were often so gradual as to be imperceptible. But a few changes were directly attributable to the depopulation caused by the epidemics. Fields which had still stood in grass when the Pilgrims arrived in 1620 were rapidly being reclaimed by forest by the time of the 1630 Puritan migrations to Massachusetts. William Wood spoke of places "where the Indians died of the plague some fourteen years ago" that were covered with "much underwood . . . because it hath not been burned." Between Wessaguscus and Plymouth, the regrowth of forest had already made one extensive area "unuseful and troublesome to travel through, insomuch that it is called ragged plain because it tears and rents the clothes of them that pass." Some Indian fields were rapidly overgrown by the strawberries and raspberries in whose abundance colonists took so much delight, but these were an old-field phenomenon that would not reproduce themselves for long without the growing conditions Indians had created for them. When the Puritan migrations began, the animals that had relied on the Indians to maintain their edge habitats were still abundant beyond English belief, but in many areas the edges were beginning to return to forest. Declining animal populations would not be noticed for many years, but habitat conditions were already shifting to produce that effect.[15]

Because Old World pathogens had such profound effects on Indian lives, any analysis of the fur trade must bear those effects constantly in mind. Changes in the ways Indians organized subsistence, made political alliances, and interacted with the environment stemmed directly from the new market in furs and trade goods, but the larger context was that of a society facing biological havoc. One historian, Calvin Martin, has gone so far as to assert that Indians became involved in the fur trade because they believed that game animals, rather than Europeans, had brought the epidemics upon them. In Martin's view, Indian demand for European trade goods was decidedly secondary to the Indians' belief that they were conducting a holy war against animals that were persecuting them with disease. Elegant as his thesis may be, it is supported by very little evidence, and none of it from New England. Martin is right to note the apparent paradox of Indian participation in the fur trade: by so willingly overhunting the beaver and other game animals, Indians across North America were responsible for attacking one of the major bases of their own subsistence. But to appeal to a spiritual holy war to explain this phenomenon trivializes the social and economic circumstances that led Indians to engage in trade in the first place. The connection between the epidemics and the fur trade was real, but neither so direct nor so purely spiritual as Martin would have us believe.[16]

The fur trade could not have existed without Indians: in order for the English to exploit beavers and other furbearers, it was essential that they have the willing cooperation of Indian partners. This fact sprang from the very hunting skills which English observers regarded as "laziness" in Indian males. Writing of the beaver, William Wood admitted that "these beasts are too cunning for the English, who seldom or never catch any of them; therefore we leave them to those skillful hunters [Indian males] whose time is not so precious." In fact, the Indians' time was not less precious: they simply hunted much more efficiently than the English—even with their supposedly "inferior" technology— and stopped when they were satisfied with what they had caught. (Perhaps their leisure time seemed to them *more* precious than the English thought their own.) No amount of English labor could have yielded so great a return on invested capital as did the Indians' labor in hunting and processing beaver. But to obtain that return, English colonists needed to offer goods that the Indians found as desirable as the Europeans found furs.[17]

Trade was nothing new to the New England Indians. "Amongst themselves," said Roger Williams, "they trade their Corne, skins, Coates, Venison, Fish, etc." Within villages, it was difficult to distinguish such trade from the gift giving that was so important in maintaining political and economic alliance networks. But trade also took place between villages, especially of goods that some possessed in greater quantities by reason of ecological circumstances. An interior village in the upland forest, for instance, which had an abundance of chestnuts, might regularly exchange them with a coastal village that had an abundance of shellfish. Northern New England Indians could obtain a much greater variety and quantity of furs than those in the south, and these could be traded for corn and beans with southern Indians who had agricultural surpluses. But even exchange between villages had important political overtones, since it served as a token of their diplomatic relations. Champlain in 1606 observed a transaction in which one sachem felt he had not received value for value: he "went away very ill-disposed towards them for not properly recognizing his presents, and with the intention of making war upon them in a short time." Most exchanges, whether internal or external to a village, were articulated in the language of gift giving.[18]

In addition to the implicit gift relationship, what distinguished this trade from that of European merchants was its preponderantly local nature. Trade took place largely between adjacent villages; no entrepreneurial class existed whose chief role was to move commodities over long distances. Goods of high value might still travel hundreds of miles, but generally only by being traded from village to village. As John Smith said, villages "have each trade with other so farre as they have society on each others frontiers, for they make no such voyages as from *Pennobscot* to *Cape Cod,* seldome to *Massachset.*" Trade was either between individuals for goods each would personally use, or between sachems for goods that could be redistributed to followers. The European fur trade could come into existence only by being assimilated into this earlier context.[19]

The objects Europeans could offer in trade had certain qualities that were completely new to Indian material culture. Brass and copper pots allowed women to cook over a fire without the risk of shattering their earthen vessels, and were much more easily transported. Woven fabrics were lighter and more colorful than animal skins and nearly as warm. Iron could be sharpened and would hold an edge better than stone, so that European hatchets and knives had advantages over Indian ones. Indians had no firearms, and were unfamiliar with alcohol. But in spite of the newness of these things, it is wrong to see the acquisition of European technology as in itself necessitating a revolution in Indian social life. European tools did not instantly increase Indian productivity in any drastic way. Most were readily incorporated into subsistence practices and trade patterns that had existed in precolonial times. They were in fact often reconverted into less utilitarian but more highly valued Indian objects: the many early explorers who came across Indians wearing brass and copper jewelry, for instance, were probably seeing what Indians believed to be—along with arrowheads—the proper use of European brass and copper kettles.[20]

Indians had first to learn the uses of European fabrics and metals before they would trade for them; as Verrazzano discovered at Narragansett Bay, this did not always happen automatically. What Indians valued was often less the inherent technical qualities of a material object than its ascriptive qualities as an object of status. (In this, they were not fundamentally different from Europeans who sought to obtain animal skins so as to display personal wealth.) A kettle or a metal arrowhead might have virtues that saved labor and were desirable in their own right, but these did not become compelling until other Indians owned them and an individual's importance began to be measured by their possession. Indians eventually sought many of the things Europeans offered in trade, not for what *Europeans* thought valuable about them, but for what those things conveyed in *Indian* schemes of value. In effect, they became different objects. Being rare and exotic, European goods could function as emblems of rank in Indian society and as gifts in the exchanges that created and maintained alliance networks. Indian individuals seeking to increase their political power, especially in the wake of the epidemics, often tried to accumulate trade goods that could be used to gain more allies. Transactions of this kind involved exchanges of values that were functionally more symbolic than utilitarian; as with the property systems we have already examined, Indians and Europeans understood their acquisitions differently, for the simple reason that those acquisitions were embedded in different social and ecological contexts.[21]

Some of the most highly desired goods offered by Europeans in fact had little to do with European technologies at all. After using up their tiny initial store of trade goods, the Plymouth colonists had, as Bradford said, "little or nothing else" to offer Indians "but this corn which themselves had raised out of the earth." Agricultural produce had been the major substance offered by southern Indians in trade with northern ones, and Plymouth at first behaved little differently from an Indian village in trading its maize for furs. What stimulated the trade was not so much new European technologies—nothing other than the sailing ship was necessary to pursue it—as a new European economic need: the need to find commodities that would repay debts to European merchants. In this sense, Europeans took hold of the traditional maize-fur trade network and transformed it from a system of binary village exchange to a link in the new Atlantic economy. Colonial governments reserved the fur trade as a sovereign right for themselves and the merchants who served as their agents, and began to amass corn both by trading with southern Indians and by taxing the colonists themselves. For a short while, it seemed as if corn raising would prove a most profitable enterprise: when Francis Higginson said that it was "almost incredible what great gaine some of our English Planters have had by our Indian Corne," he was excited not about the grain itself but about its easy convertibility into furs. Transported in English ships from southern New England to the Maine coast, corn remained one element of the fur trade throughout the seventeenth century, but its importance nevertheless declined. Its disadvantages limited its desirability as a staple commodity: it was bulky, relatively difficult to transport, and its value fluctuated considerably relative to both the size of the annual corn crop and the northern Indians' success in hunting for food.[22]

It was another commodity—like maize, more Indian than European—that revolutionized the New England fur trade: wampumpeag, the strings of white and purple beads we know today as wampum. Made by grinding and drilling the shells of whelks and quahogs until they were hollow cylinders a quarter of an inch long and an eighth of an inch in diameter, its manufacture was ecologically limited to the Long Island Sound area where these shellfish flourished. Never made in great quantities during precolonial times, wampum was a highly valued token of personal power and wealth. It was initially rare outside the coastal villages in which it was made, so that elsewhere, as Bradford noted, only "the sachems and some special persons . . . wore a little of it for ornament." Lesser individuals dared not accumulate too much of it unless they were willing to challenge those with higher prestige. It was exchanged mainly at well-circumscribed ritual moments: in the payment of tribute between sachems, as recompense for murder or other serious crimes, in the transfer of bride-wealth when proposing marriage, as payment for a powwow's magical services, or in gift exchanges to betoken friendship and alliance. It was, in other words, a medium of gift giving whose value was widely accepted among the Indians of southern New England.[23]

To Europeans, wampum was ideally suited to become the medium for a wider, more commercial exchange—to become what John Locke called "money." The Indians' adoption of European metal drills increased their production of wampum, making it more widely available for trade. But, as with corn, the chief European innovation was to introduce a new functional role into Indian economies, that of the merchant who transported goods between communities which, for cultural and ecological reasons, valued those goods differently. The Dutch first discovered wampum's value in 1622, and were astonished at how much it facilitated their trade. Fearing that the Pilgrims would become rivals in the Connecticut fur trade if they found out about wampum independently, the Dutch West India Company's agent, Isaack de Rasieres, decided to introduce them to it himself. He accordingly sold them £50 worth of it in 1627 and encouraged them to try trading it on the Maine coast (rather than on the Connecticut, where the Dutch trading houses were located). Although Maine Indians were initially reluctant to acquire wampum, within two years it had become the single most important commodity Plymouth had to offer. Presumably because of the high prestige that Indians associated with its possession, Plymouth traders "could scarce ever get enough for them, for many years together," and were able to cut off other traders who lacked wampum and had only European goods to sell.[24]

Control of wampum rapidly became crucial to both Indians and Europeans. The political crisis created by the epidemics made wampum a necessary acquisition for any sachem trying to expand—or even retain—his or her power. As greater and greater quantities became available to more and more individuals, even to those who had once been of low rank, an inflationary cycle in the price of prestige objects fueled trade all the more. Possession of wampum became increasingly common, with widening effects on status systems. Here again, wampum was part of the reorganization of Indian economic and political life which followed in the wake of the epidemics: competition for its acquisition established new leaders, promoted dependence on European traders, and helped shift the tribute obligations which had previously existed among Indian villages. "Strange it was," said Bradford, "to see the great alteration it made in a few years among the Indians themselves."[25]

On the European side, the importance of wampum to the fur trade made it imperative that colonies have a guaranteed supply of it. That meant controlling trade with the Indians of Long Island Sound—Pequots, Mohegans, Narragansetts, and villages on Long Island—who could procure the necessary whelks and quahogs. At first their wampum was obtained by flooding them with high-status European goods such as kettles and firearms; in response, they expanded their production of it and, as Bradford said, "became rich and potent by it." The number of Europeans trying to trade with the wampum makers encouraged the Indians to become shrewd bargainers who played one trader off against another. According to Roger Williams, they "beate all markets and try all places, and runne twenty thirty, yea forty mile, and more, and lodge in the Woods, to save six pence." In part because of the guns which these markets gave the Indians of the south coast, many colonists increasingly feared the Indians' power and sought a less dangerous way of acquiring their wampum. Governor Bradford went so far as to write a poem about the problem:

> But now they know their advantage so well,
> And will not stick, to some, the same to tell,
> That now they can, when they please or will,
> The English drive away, or else them kill.

The colonists' economic problem of obtaining a sure supply of wampum and the military problem of dealing with independent Indian arms were finally solved simultaneously by means of armed force: the slaughter of the Pequots in 1637 and the assassination of the Narragansett sachem Miantonomo in 1643. Exacting a regular military tribute in wampum proved a safer and more reliable source of supply than trading guns for it.[26]

The fur trade was thus far more complicated than a simple exchange of European metal goods for Indian beaver skins. It revolutionized Indian economies less by its new technology than by its new commercialism, at once utilizing and subverting Indian trade patterns to extend European mercantile ones. European merchants created an expanded regional economy in New England by shuttling between several different trading partners: wampum producers along Long Island Sound, corn growers— both Indians and colonists—in the south generally, European manufacturers, and the Indians—located primarily in the north —who hunted furs. Trade linked these groups with an abstract set of equivalent values measured in pelts, bushels of corn, fathoms of wampum, and price movements in sterling on London markets. The essential lesson for the Indians was that certain things began to have *prices* that had not had them before. In particular, one could buy personal prestige by killing animals and exchanging their skins for wampum or high-status European goods.[27]

Formerly, there had been little incentive for Indians to kill more than a fixed number of animals. As Nicolas Denys observed of the northern Indians, "they killed animals only in proportion as they had need of them. They never made an accumulation of skins of Moose, Beaver, Otter, or others, but only so far as they needed them for personal use." The one occasion for which furs *were* accumulated in precolonial times—when they were exchanged with southern villages for corn and other goods—was in fact an exception to prove the rule. Aside from limitations on the amount of grain Indians could move overland or by canoe, such precolonial trade was kept to modest levels by two factors. Need —as measured by use and by the success of harvest or hunt—still determined the volume of trade: there was little reason for the inhabitants of a village to trade for more food than they could eat, or more clothes than they could wear. Taking place mainly between villages, the trade was conducted by sachems and other high-prestige individuals, and so was held in check by the status relationships within a village and the diplomatic relations between villages. It had few of the expansionist tendencies of European commerce.[28]

Precolonial trade enforced an unintentional conservation of animal populations, a conservation which was less the result of an enlightened ecological sensibility than of the Indians' limited social definition of "need." One Indian at the end of the eighteenth century remembered that in earlier times his people had not killed "more than necessary." He said this was because "there was none to barter with them that would have tempted them to waste their animals, as they did after the Chuckopek or white people came on this island, consequently game was never diminished." As we have seen, European trade changed all of this. It introduced Indians to a new set of prestige goods which could only be obtained by trade; moreover, the disruption of earlier status systems by the epidemics eliminated many of the social sanctions which had formerly restricted individual accumulation. Indian economies thus became attached to international markets, not for technological, but for social and political reasons. For them, it was a market of much more limited circulation than it was for Europeans: Indian notions of status were measured by a handful of goods, whereas Europeans could accumulate wealth with virtually any material possession. Nevertheless, even a limited market in prestige was enough to turn Indians into the leading assailants of New England's furbearing mammals.

Certain animal populations began to decline in consequence, even though the epidemics meant that fewer Indians were hunting them. The commercialization of the Indians' earlier material culture thus brought with it a disintegration of their earlier ecological practices.[29]

Chief among the animals which suffered from the fur trade was of course the beaver, whose low reproductive rates and sedentary habits made it easily threatened by concentrated hunting. Never abundant in southern New England, it was disappearing from Massachusetts coastal regions by 1640 and had ceased to be of much economic significance in the Narragansett country by 1660. The southern trade lasted longest on major rivers that drained extensive regions to the north, where Indian hunters continued to find beavers to kill. On the Connecticut River, European traders established in turn the towns of Wethersfield, Hartford, Windsor, and Springfield as fur posts. Each successive one captured Indian trade from those below it, and the arrival of agricultural settlers eventually brought the fur trade of each to an end. Springfield, established by William Pynchon in 1636, maintained its hold longer than most. Although its trade had declined by 1650 to the point that Edward Johnson could describe it as of "little worth," Pynchon's son John managed between 1652 and 1658 to procure from Indians nearly 9000 beaver pelts, in addition to hundreds of moose, otter, muskrat, fox, raccoon, mink, marten, and lynx skins. Output gradually declined, experienced a sharp drop in the 1670s during the conflicts surrounding King Philip's War, and from then on continued only at much reduced levels. By the end of the century, the fur trade had lost its economic importance to the area. The same fate overtook the string of trading posts—Concord, Chelmsford, and Lancaster—which were established on the Merrimac River.[30]

Other southern animals were also at risk. As Pynchon's records showed, virtually any furbearer could be sold for its pelt. But in the vicinity of denser English settlements, especially Boston, there was also a market for meat, one in which both Indians and colonists participated. In the early 1630s, a male turkey weighing forty pounds brought four shillings in eastern Massachusetts, while other birds—heathcocks, ducks, geese, and so on —cost from four- to sixpence. In Springfield, Pynchon bought venison from the Indians by paying wampum for it. To increase colonial meat supplies, the Massachusetts General Court gave exclusive hunting privileges to designated colonists who improved a pond by setting decoys on it, or who used nets to hunt birds on certain islands. Other colonists used Indian servants to hunt food for them. For their part, Indians found colonial guns particularly helpful in hunting birds and large mammals. They sold enough of what they killed for Governor Bradford to object that

> The gain hereof to make they know so well,
> The fowl to kill, and us the feathers sell.
> For us to seek for deer it doth not boot,
> Since now with guns themselves at them can shoot.
> That garbage, of which we no use did make,
> They have been glad to gather up and take;
> But now they can themselves fully supply,
> And the English of them are glad to buy.

Bradford's main anxiety was that Indians owned guns at all, but he was also irritated by their killing and selling things which English colonists might otherwise have obtained for nothing.[31]

Overhunting combined with reductions in edge habitats led some of the meat species to decline in numbers even by the end of the seventeenth century. Already in 1672, John Josselyn said of the turkey that English and Indian hunters had "now destroyed the breed, so that 'tis very rare to meet with a wild turkie in the woods." Only domesticated ones could be seen in eastern Massachusetts. A century later, the bird which William Wood had seen in flocks of a hundred had become so rare throughout New England that a popular farmer's manual could define the word "turkey" as "a large domestick fowl, brought from Turkey, and is called by the name of its country." The fact that the domesticated bird of Europe had originally been brought from America by Spanish colonists in the sixteenth century had apparently been forgotten. Other birds would eventually be eliminated as well—the passenger pigeon, which had existed in so many "millions of millions," began to disappear toward the end of the eighteenth century—but the fate of these other species was not finally sealed until the rise of metropolitan markets for fowl in the nineteenth century.[32]

More important to southern Indians was the gradual disappearance of the white-tailed deer and other large herbivores. Deer were threatened by changes in their habitat, augmented numbers of hunters, and competition from domestic livestock. They were so reduced by the end of the seventeenth century that Massachusetts enforced its first closed season on their hunting in 1694, and in 1718 all hunting of them was forbidden for a closed term of three years. By the 1740s, a series of "deer reeves"—early game wardens—were regulating the deer hunt, but to little avail. At the end of the eighteenth century, Timothy Dwight noted that deer were "scarcely known below the forty-fourth degree of north latitude," having vanished from all but the northern stretches of Vermont, New Hampshire, and Maine. With them, save for in the far north, had gone the elk, bear, and lynx. "Hunting with us," said Dwight, "exists chiefly in the tales of other times."[33]

By the time this happened, of course, the colonists no longer relied on hunting for any significant portion of their subsistence. The real losers were the Indians, whose earlier way of life was encountering increasing ecological constraints. Indian settlement patterns began to change in the 1630s. On the south coast, Indians took to occupying coastal sites year-round in order to stockpile shellfish so that they could make wampum on an expanded scale. This new sedentarism was reinforced, and promoted elsewhere in New England, by military conflicts—themselves a consequence of the political turmoil which followed the epidemics and expanding European trade—that led villages to prefer permanently fortified sites inhabited by relatively dense populations. Attacks by colonists and intertribal warfare concentrated Indians with particularly dramatic consequences: Gookin described how it forced "many of them to get together in forts; by which means they were brought to such straits and poverty, that had it not been for relief they had from the English, in compensation for labour, doubtless many of them had suffered famine." But even when the threat of violence was not so immediate, Indians were living in fixed locations on a more permanent basis. Earlier subsistence practices which had depended on seasonal dispersal were gradually being abandoned, with important social and ecological effects. Denser settlement patterns encouraged the spread of infectious diseases and increased pressure on adjacent hunting and planting areas. As a result, Indians found themselves relying on a narrowing range of foodstuffs.[34]

As Indians increasingly sold the skins they hunted, they had to have an alternative for their own clothing. Despite some initial reluctance, they found it in European fabric, which, next to wampum, was by mid-century the single most important commodity they bought with fur. At Pynchon's trading post in the 1650s, Indian transactions for textiles outnumbered transactions for metal goods more than fivefold. Elsewhere, the price in beaver skins for a blanket was identical to that for a gun. European clothing was not only high in prestige value but cheaper than fur to own: a large beaver pelt cost nearly twice as much in wampum as a finished cloth coat. Raw cloth, which Indians preferred, was even cheaper. By the 1670s, Daniel Gookin could write:

> The Indians' clothing in former times was of the same matter as Adam's was, viz. skins of beasts, as deer, moose, beaver, otters, raccoons, foxes, and other wild creatures. . . . But, for the most part, they sell the skins and furs to the English, Dutch, and French, and buy of them for clothing a kind of cloth, called duffils, or trucking cloth, about a yard and a half wide.

From duffils they learned to fashion their clothing, and eventually they had no choice but to do so. The decline in deer populations made their reliance on European fabrics inescapable.[35]

By the 1660s, southern New England Indians had for several decades been relying on wampum and furs to trade for textiles, arrowheads, knives, guns, and kettles. Gookin reported that "they generally disuse their former weapons." During the seventeenth century, earlier forms of these implements lost prestige value and young Indians growing up gained less and less experience in making them. Villages thus became vulnerable to changes in their trade base in a way they had not been before. Wampum, which in the 1640s had circulated as legal tender among colonists as well as Indians, lost much of its value in the 1660s when European demand for beaver declined and new supplies of silver coin from the West Indies reduced the scarcity of colonial specie. As a result, although wampum continued to circulate in the Indian trade, the colonists no longer defined it legally as money.

"It is but a commodity," declared the Providence Court, and "it is unreasonable that it should be forced upon any man." Demand for wampum fell, and Indians on the south coast suddenly found themselves isolated from markets on which they had come to rely. Indians for whom pelts had been their main access to trade had comparable experiences when their fur supplies gave out. These changes contributed to the conflicts leading up to King Philip's War in 1675–76, but their longer-term effect was to force Indians who depended on trade goods to turn to the only major commodity they had left: their land.[36]

The second half of the seventeenth century saw Indians in southern New England lose most of their land. The colonists accomplished this dispossession in ways which have been recounted at length elsewhere and need not be repeated here. Whether seized as the spoils of war, stolen by colonial subterfuge, or simply sold by Indians to obtain trade goods, the net effect was the same: decreasing quantities of land remained free for the Indians' use. The denser settlement patterns they had adopted earlier became less and less a matter of choice as they found themselves more and more surrounded by colonists. Subsistence practices which had never before had deleterious ecological consequences began gradually to have them. Planting fields could no longer be so easily abandoned when their fertility declined, and agricultural yields fell, making crops a less reliable source of food. Hunting too became more difficult. Adjacent colonial settlements eventually tried to restrict Indian hunting on English land, and such key food sources as deer became harder to obtain. Toward the end of the seventeenth century, many Indians were actually beginning to keep European livestock. If the evidence from one archaeological site in Connecticut can be generalized, by 1700 some Indian villages were relying on domesticated animals for over half their mammalian meat supply. The keeping of cattle on Indian land further decreased the forage available for wild deer herds and so continued the erosion of hunting resources. From an ecological point of view, Indian subsistence practices were in many ways more and more like those of European peasants. As European trade had done earlier, European agriculture reorganized Indian relationships within both the New England regional economy and the New England ecosystem.[37]

Many of the changes I have been describing apply primarily to southern New England. In the north, where Indian populations were much smaller in relation to their land base and fewer colonists came to settle, pressure on animal populations resulted almost wholly from trade rather than competition over land. Low Indian densities meant fewer hunters and for that reason larger concentrations of the very animals Europeans most desired, so that the fur trade was far more active in Maine and eastern Canada than it was farther south. In response, Indians extended their hunting of furbearers to seasons when those animals had not traditionally been taken, and winter subsistence activities which had earlier relied on a wide range of species turned increasingly to the handful that had become tradeable commodities. When Indians returned to the coast in summer to sell their catch, they sought not only wampum and the preferred European tools, but food as well, including corn, bread, peas, prunes, and alcohol. Because of their larger and more concentrated involvement with the fur trade, their eventual dependence on it was even greater than that of southern New England Indians. Nicolas Denys, writing in 1672, said that they had "abandoned all their own utensils." In effect, furbearers were being asked to supply not only clothing and winter food but many other aspects of subsistence as well.[38]

As in the case of southern New England Indians, certain forms of European technology became integrated into northern subsistence practices, with important effects. Once Indians stopped cutting up brass and copper kettles for jewelry, they were liberated from the large and cumbersome wooden kettles, made from tree stumps, which they had once used for cooking. Denys said that the earlier "immovable kettles were the chief regulators of their lives, since they were able to live only in places where these were." European kettles enabled villages to follow animal populations more readily and without the necessity of resorting to base camps; to Indians, they seemed "the most valuable article they can obtain from us." Likewise, the musket, although rarely used for hunting beaver or the smaller furbearers because of its inaccuracy, made the killing of large herbivores a simpler task. Shooting a moose provided far more food with much less labor than killing many birds or small game animals. A solitary beast which had once been hunted only when deep snows slowed its movements, the moose suddenly became an easier prey. As a result, it was gone from parts of eastern Canada by the mid-seventeenth century, and the same pattern repeated itself in northern New England. At the end of the eighteenth century, the colonial historian James Sullivan said of Maine that "the moose, a monstrous large animal, has been plenty there in former days, but it is rare to see one at the present time."[39]

What encouraged the destruction of the beaver, on the other hand, was less new technologies than new market relations. Because supplies of the animal lasted much longer in the north than in the south, northern Indians developed a new set of institutions for controlling the hunt. Reduction in beaver populations, as well as conflict over who should be permitted to trap where, led to major shifts in Indian notions of property. Northern villages had formerly divided their hunting territories on a shifting and ad hoc basis; now, as families tried to hang on to their share of a declining beaver hunt, such territories became more and more fixed. By the eighteenth century, Maine Indians had allocated their lands into family hunting territories whose possession was inherited from generation to generation. In 1764, the fur trader Joseph Chadwick wrote that "their hunting ground and streams were all parcelled out to certain families, time out of mind." By turning uncaught beavers into private property, Indian families sought to guarantee their conservation: "it was their rule," said Chadwick, "to hunt every third year and kill two-thirds of the beaver, leaving the other third part to breed." Indians had once conserved their game animals, probably without fully realizing it, by their seasonal rotation from habitat to habitat, moving wherever food could be had with least effort. Now their dependence on trade forced them to control the beaver hunt much more self-consciously in order to assure that there would be animals to bring to market. The beaver, in other words, had ceased to be an object of use, conserved because the need for it was slight; instead, it had become a commodity of exchange, conserved because the need for it was great.[40]

In part because of the family hunting territories, the beaver survived in Maine, albeit in much reduced numbers. The Maine fur trade continued throughout the eighteenth century, but the effects of colonial warfare, overhunting by English trappers, and competition from Canadian furs combined to make it less and less profitable. At the end of the French and Indian Wars, Maine Indians were complaining that "since the late war English hunters kill all the Beaver they find on said streams, which had not only empoverished many Indian families, but destroyed the breed of Beavers." Elsewhere in New England, the beaver barely survived into the nineteenth century; the southern fur trade was no longer very profitable by 1700, and the animal itself remained only in isolated areas. By 1797, Benjamin Trumbull could write of Connecticut's beavers—as well as its otters, foxes, martens, raccoons, minks and muskrats—using the past tense. Samuel Williams, in his *History of Vermont,* said that "the beaver has deserted all the southern parts of Vermont, and is now found only in the most northern, and uncultivated parts of the state." Even there, the animal would be gone before long.[41]

The elimination of the beaver had ecological consequences beyond the loss of the animals themselves. Their passing left New England with a wealth of place names which no longer made much sense: scattered across the map of the region one still finds Beaver Brooks, Beaver Stations, Beaver Creeks, and Beaver Ponds. More importantly, they left behind a series of artifacts that would await European settlers when they came to clear the land. When colonists cut a road through freshly opened countryside, they often went out of their way to cross streams on abandoned beaver dams rather than to build bridges. The sites of old dams were often chosen as preferred mill sites. Some beaver ponds became spawning grounds for shad and salmon, thus providing sites where fish for food and fertilizer could be had with a minimum of labor. But it was when the old dams collapsed for want of maintenance that they conferred their greatest benefit on colonial settlers. Behind them was many years' accumulation of leaves, bark, rotten wood, and rain-washed silt; in addition, their ponds had killed acres of trees which had once stood on the banks of pre-beaver streams. When the pond disappeared with the breaching of its dam, the rich black soil was suddenly exposed to the sun and rapidly became covered with grass that grew "as high as a man's shoulders." Not only did this provide forage for moose and deer—as long as those animals remained to browse there— but it became ideal mowing ground when settlers arrived with their cattle. As the English traveler Henry Wansy remarked, "It is a fortunate circumstance to have purchased lands where these industrious animals have made a settlement. At some of them, there have been four ton of hay cut on an acre." The old pond bottoms, which could be as much as two hundred acres in extent, provided excellent agricultural land as well. As one colonial writer put it, "Without these natural meadows, many settlements could not possibly have been made, at the time they were made." The death of the beaver in fact paved the way for the non-Indian communities that would soon arrive.[42]

By 1800, the joint efforts of Indians and colonists had decimated many of the animals whose abundance had most astonished early European visitors to New England. Timothy Dwight in the early nineteenth century said of Connecticut that "we have hardly any wild animals remaining besides a few small species of no consequence except for their fur." Such animals had fallen victim especially to the new Indian dependence on a market in prestige goods. The Indians, not realizing the full ramifications of what that market meant, and finally having little choice but to participate in it, fell victims too: to disease, demographic collapse, economic dependency, and the loss of a world of ecological relationships they could never find again. No one understood this better than the Indians themselves. In 1789, the Mohegans petitioned the state of Connecticut for assistance, explaining:

> The times are Exceedingly Alter'd, Yea the times have turn'd everything upside down, or rather we have Chang'd the good Times, Chiefly by the help of the White People, for in Times past, our Fore-Fathers lived in Peace, Love, and great harmony, and had everything in Great plenty. . . . But alas, it is not so now, all our Fishing, Hunting and Fowling is entirely gone.

Even if they exaggerated the peace, love, and harmony of precolonial Indian life, they did not mistake its plenty. Indian economies had maintained and relied on that plenty, and could not exist without the ecological relationships it implied. But although selling the animals had been the Indians' major contribution to their new circumstances, it was by no means the only reason their world had turned upside down. Ecological changes wrought by the colonists themselves were far more extensive, and needed no Indian partners for their accomplishment.[43]

6

TAKING THE FOREST

The furbearers of colonial New England had been destroyed in two ways: by having a price placed on their heads and by losing their ecological habitats to new human uses of the land. As we have seen, the edge habitats once maintained by Indian fires tended to return to forest as Indian populations declined. But edge environments were also modified or reduced—and on a much larger scale—by clearing, an activity to which English settlers, with their fixed property boundaries, devoted far more concentrated attention than had the Indians. Whether edges became forests or fields, the eventual consequences were the same: to reduce—or sometimes, as with European livestock, to replace —the animal populations that had once inhabited them. The disappearance of deer, turkey, and other animals thus betokened not merely a new hunting economy but a new forest ecology as well.

Colonists cut down trees for many reasons. Some of these—like clearing fields for agriculture—were a necessary adjunct of a European rural economy, and often bore only an indirect relationship to production for market. Others, such as lumbering, were much more immediately involved with mercantile activity and trade. Like furs, timber products were among the earliest "merchantable commodities" colonists sent back to Europe to repay their debts to financial backers. In 1621, when the Pilgrims made their first shipment home in the 55-ton vessel *Fortune,* they sent only two barrels of furs; the rest of the ship's hold was, as Bradford reported, "laden with good clapboard as full as she could stow." Even more than furs, whose acquisition required an exchange of trade goods with their Indian hunters, timber was free for the taking. One had theoretically to own the land on which it stood, but this was an easy rule to evade. Much of the value inhering in timber appeared to be a gift of nature, requiring only a modest investment of labor and capital to be turned into profit. To "improve" timber, and so acquire sure property rights to it, one had merely to cut it down, saw or split it into manageable sizes, and—the most costly single step—ship it to market. In some areas, this was done in conjunction with clearing for agricultural settlement; in others, the cutting of timber was a chief economic activity in its own right.[1]

Colonists sought different species of trees for different purposes, and so, when lumbering as opposed to clearing, cut forests selectively. White oak was the chief wood used for the timbers and planking of ships, and proved to be ideal for barrel staves as well. Black oak, which was less suitable than white for most maritime uses, was appropriate for the underwater portions of a ship's timbers because of its resistance to boring by tropical worms. Cedars and chestnuts were also relatively immune to rot and well suited to use in exposed outdoor sites. Pitch pine furnished a wide range of naval stores, including pitch, turpentine, and rosin. Most dramatic of all, the white pine, towering above all other trees in New England forests, was a perfect source of ship's masts. The wide range of uses these trees served in a maritime trade economy meant that the demand for each varied, and that each to some extent was traded in a different market. It also meant that the habitats in which they grew were subject to different stresses by colonial lumbering activities.[2]

From the 1630s onward, the largest concentration of commercial lumbering for export was located in Maine and New Hampshire along the major rivers north of the Merrimac. There, on the sites of old forest fires, stood tracts of white pine containing trees as much as four to six feet in diameter, and 120 to 200 feet in height. Trees of such size and straightness were unknown in Europe; no European trees were large enough to serve for masts without several being spliced together. Up until the development of lumbering in America, England had been forced to rely for its ships on the pieced-together masts of Scotch fir which it received in trade from Baltic forests. With the closing of Baltic trade during the Dutch war of 1654, the Royal Navy commissioners began to look to the white pines of New England for an alternate mast supply. The first sawmills in Maine had been established in the early 1630s, but it was not until after mid-century that the northern lumbering industry came into its own. By 1682, there were twenty-four sawmills operating at the sites of present-day Kittery, Wells, and Portland, shipping principally softwood lumber which, unlike hardwood, would float on the navigable streams flowing down to the coast. The mouth of the Piscataqua rapidly became the chief lumber port of the northern colonies: as Samuel Maverick remarked in 1660, "Most of the Masts which have come for England" had been gotten on that river. Some of these were transported in special mast ships of a length and tonnage suited to carrying two or three dozen whole pine trees for use by the Royal Navy as masts. New England forests thus became a key resource for maintaining English naval power.[3]

England's experiences with timber and fuel shortages, as well as its involvement in European wars that threatened to cut off its foreign supply of shipbuilding materials, led its rulers to adopt what seemed to many colonists an overly cautious attitude toward conserving New England's forests. When Massachusetts received its new charter in 1691, a clause was inserted that reserved for the Royal Navy "all [mast] trees of the diameter of twenty-four inches and upwards at twelve inches from the ground." A fine of £100 per tree was to be assessed against any person "Felling Cutting or Destroying" without royal license any such trees not already in private hands. The terms of the charter were eventually supplemented by additional restrictions. In 1704, royal surveyors were charged with marking all potential mast trees with a blaze shaped like a broad arrow to ensure the Crown's right to them. At the same time, pitch pines were protected to maintain the Navy's supply of pitch and turpentine. These regulations were repeatedly renewed until the American

Revolution; if enforced, they might conceivably have slowed the cutting of New England forests. Enforcement proved virtually impossible, however, and so the "broad-arrow laws" came to stand more than anything else for European anxieties about deforestation. The colonists violated the laws constantly.[4]

New England lumbering used forests as if they would last forever. Because prime mast trees were usually scattered among those of lesser value, many less-than-perfect trees were simply destroyed when larger ones were felled. Colonists were usually far more interested in conserving their own labor than in using available timber resources to the full. A favorite labor-saving device was the technique known as "driving a piece," in which lumberers cut notches in a row of small trees and then felled a larger tree on top of them, thus cushioning its fall so as to protect it from shattering. The method was ideal if an area was to be used for farming once lumber trees were removed, since it brought down many more trees with much less work. Any trees without market value could simply be burned where they fell. Waste was compounded by high colonial standards of what was and was not marketable lumber. Even purposes for which Europeans made do with low-quality wood were accomplished by New Englanders with the finest available lumber. A memoir written in the Androscoggin River country in 1800 recalled how

> The richest and straightest trees were reserved for the frames of the new houses; shingles were rived from the clearest pine; baskets, chair bottoms, cattle bows, etc., were made from brown ash butts; all the rest of the timber cleared was piled and burned on the spot. . . . All the pine went first. Nothing else was fit for building purposes in those days. Tables were made 2 1/2 feet wide from a single board, without knot or blemish.

This was a pattern that would characterize American lumbering until the late nineteenth century.[5]

Not only the white pine suffered from these practices. Although New England white oak was shunned by English shipbuilders as "tender" and too easily prone to decay, colonial carpenters were not so fastidious. They used it heavily. Moreover, sugar plantations in the West Indies and wineries in the Madeiras needed barrels in which to ship their commodities to European markets. Their voracious consumption of wood in burning fuel and clearing land left many of the Atlantic and Caribbean islands completely deforested, so that by the late seventeenth century they were dependent on New England for staves and hoops made of red as well as white oak. In New England, colonial tanners used increasing quantities of oak bark as their operations grew, and oak lumber was also favored for building construction. Because of their lightness and resistance to decay, white and red cedar were considered ideal woods for the manufacture of shingles, clapboards, and fence posts. Colonial architecture came to rely on thin-walled structures supporting relatively lightweight roofs, and so cities like Boston developed significant markets in cedar shingles. Even some Indians were drawn into the trade. Daniel Gookin told of one Indian village whose inhabitants earned "many a pound, by cutting and preparing cedar shingles and clapboards, which sell well at Boston and other English towns adjacent."[6]

Such uses of wood placed prices on certain New England trees just as the fur trade had placed prices on New England mammals: the forest itself came to have a value at market. Neither white pine nor white cedar—extensive stands of the one being limited to dry ridge tops and old burn sites, and of the other to streambanks and wet swamps—had ever been abundant in New England. Timothy Dwight gave an indication of their relative numbers when he said that if the white pines south of Maine "were all collected into one spot," they "would scarcely cover the county of Hampshire" in England; the white cedars, still rarer, "would scarcely fill three townships." Neither tended to regrow when cut—pine was generally replaced by hardwoods, and cedar by red maple—so that the populations of both were easily reduced by lumbering activities.[7]

Speaking of the white pines he had once seen near Lebanon, New Hampshire, Dwight wrote that "a great part of them have since been cut down." To show that this was not merely a local effect, he added, "There is reason to fear that this noblest of all vegetable productions will be unknown in its proper size and splendor to the future inhabitants of New England." His prophecy soon came true, but full removal of the original northern pines did not come until the lumbering boom of the nineteenth century. In coastal regions and in southern New England, they disappeared from many areas much earlier. When the West Indian merchant James Birket visited the Piscataqua country in 1750, he found its inhabitants complaining that their lumber was "far to fetch out of the Country and Stand[s] them very dear"— this in the region which a century before had been the center of northern lumbering. All along the main coastal road, Birket noted, the land was "generally Cleared." Even at Dover, miles upstream, he learned that lumber was "very dear to them the Cartage out of the woods to the mills being now a long way." Lumbering continued, but moved relentlessly upstream, extending the local areas from which trees had been removed. When Dwight visited the region half a century later, he reported that "the forests are not only cut down, but there appears little reason to hope that they will ever grow again." Although the soil seemed productive where crops were planted, he said, "From the extensive nakedness which meets the eye, it is difficult for the imagination not to pronounce it barren."[8]

Although colonial writers tended to notice the disappearance of pine first, probably because of its high value and visibility, other species also suffered from wasteful cutting. One of these was cedar. What the Swedish naturalist Peter Kalm wrote of the New York region in the middle of the eighteenth century was equally true of New England: "Many swamps are already quite destitute of cedars, having only young shoots left." Heavy use was "not only lessening the number of these trees, but . . . even extirpating them entirely." As in the case of beaver, what remained behind were place-names that no longer made much sense. The historian Peter Whitney mentioned a stream in Worcester, Massachusetts, which "was formerly called Cedar Brook," although cedars no longer existed in the town through which it flowed; a similar fate awaited many of New England's various Cedar Swamps. White oaks, which did not share the reproductive problems of pines and cedars, were in no danger of disappearing from New England forests. But their usefulness in all facets of construction meant that larger trees, which furnished the best lumber, were becoming relatively more scarce in many areas by the end of the colonial period. Hickories, which colonists preferred for fuel, also became less abundant. Colonial cutting reduced not only the numbers but the sizes of these trees. Dwight reported that many Europeans who visited New England in the late eighteenth century were surprised at the small girth of its trees, a fact which many of them attributed to "sterility of soil." In fact, Dwight said, "the real cause was the age of the trees, almost all of which are young." Older trees were gone, and most of those remaining were of second growth.[9]

As early as the 1790s, proposals began to be aired suggesting that forest preserves be created to protect their timber. The Revolutionary War general Benjamin Lincoln even suggested that programs be instituted to promote the planting of acorns. This was necessary, he said, because

> our timber trees are greatly reduced, and quite gone in many parts. In towns near and bordering on the sea shore, little can now be found within the distance of twenty miles; and it is not uncommon for the builder to send at this day from thirty to forty miles for timber and planks, and the stock fast decreasing, not only from the demand of timber and planks, but from scarcity of other fuel.

In a time when all heavy goods had to move either by boat on navigable streams or by horsecart on bad dirt roads, the economic cost of transporting lumber thirty or forty miles was not trivial. That it could be gotten at all, however, demonstrates the local extent of the deforestation produced by colonial lumbering.[10]

Perhaps surprisingly, the lumberer was not the chief agent in destroying New England's forests; the farmer was. The earliest settlements had tended to be established on land that was already cleared, whether by Indians, by the departed beavers, or by annual river floods. (Flooded lands, among the richest sites for agriculture, were the *intervals* so favored by colonial farmers.) The first mowing grounds were salt marshes and sedge meadows, likewise cleared of trees. Nevertheless, it was not long before settlement pressed out into the forest itself, and here farmers encountered the problem of removing trees—not just selected species, but all of them.

As with lumbering, some habitats were more subject to such clearing than others. Colonial farmers quickly learned that certain tree species were associated with certain kinds of soil, some of which were better than others for agricultural crops. As Jedidiah Morse explained in his *American Geography*,

> Each tract of different soil is distinguished by its peculiar vegetation, and is pronounced good, middling, or bad, from the species of trees which it produces; and one species generally predominating in each soil, has originated the descriptive names of oak land—birch, beach, and chesnut lands—pine barren—maple, ash, and cedar swamps, as each species happen to predominate.

Trees that required and maintained moist forest conditions, such as hickories, maples, ashes, and beeches, generally produced a rich black humus beneath their fallen leaves, and settlers interpreted them as indicators of prime agricultural land. Oaks and chestnuts, with their denser undergrowth and more frequent groundfires, had thinner soils which required more work before they would produce favorable European crops. Still less desirable were the acidic and often sandy soils beneath various conifers—moist under hemlocks and spruces, dry under pitch and white pines—and colonial farmers avoided these wherever they could. Worst of all were areas where thorny bushes had taken over abandoned Indian fields, or where dense tangles of scrub vegetation covered lowland swamps. Farming in these places was rarely worth the bother.[11]

Colonial observers like Morse were at least partially misled when they attributed the tree species of a district to its preexisting soils: forests caused soils as much as soils caused forests. The relative fertilities of various lands in part resulted from inherent physical properties of the soil, but also from processes maintained over long periods of time by the forest itself. Trees affected soil through a multitude of mechanisms: the spread of their root systems, the amount of light they allowed to reach the forest floor, the quantity of water which they lost by evaporation from their leaves, their susceptibility to fire, the chemistry and quantity of their annual leaf fall, and so on. The net effect of these mechanisms was to make the forest an astonishingly efficient system for capturing, concentrating, and retaining nutrients from rainwater and other sources. Most soil in a forest was there because the forest kept it there. This being the case, soils changed when their parent forests were removed. As Dwight noted, newly cleared lands were "uniformly valued beyond their real worth" because their soils were blessed with a temporary gift of fertility from a forest which was no more. It was just this fact that had led Indians to practice a shifting agriculture, and as the next chapter will show, it was one that had important consequences for colonial farmers as well.[12]

At least two different methods were used in colonial times to clear forests for agriculture. Which was chosen depended on the nature of the forest, the amount of labor available, the wealth of the farmer, and the local market demand for wood. Once trees had been cut for such immediate needs as fencing and house building, early settlers tended to use the simplest and least labor-intensive technique for destroying the rest: *girdling*. Bark was stripped in an encircling band from each of the larger trees, and grain, generally maize, was planted Indian-style in mounds beneath them. Removing their bark prevented trees from leafing and eventually killed them, thus allowing enough light to reach the ground for crops to grow. Undergrowth was burned in early spring to suppress the original vegetation, and trees were removed as they eventually rotted. At the end of several years, a cleared field was the final result.[13]

Girdling had advantages and disadvantages. It wasted large amounts of wood, but by allowing trees to stand for several years before they were removed, it returned at least some of their nutrients to the soil. Letting trees rot thus helped make soil nutrients available for a relatively extended period of time. When the dead trees finally fell, however, they tore up large amounts of soil with their roots, creating pits in fields and sometimes leaving the pitted areas soilless. Worse still, rotting trees had a disconcerting tendency to drop branches and fall over at unpredictable times, often damaging crops and fences and occasionally killing livestock and people. The job of clearing was left unfinished for a number of years, since the living roots had to be grubbed out before European-style plowing was possible. At a minimum, rotting trees eventually became a serious nuisance and were regarded by European travelers as having what Dwight called "an uncouth and disgusting" appearance. Girdling thus had the virtues of saving labor and conserving soil nutrients, but these were purchased by wasting wood, running risks of accident, and devoting long years to the incomplete removal of trees and roots.[14]

For all these reasons, an alternative method of clearing gradually became almost universal by the second half of the eighteenth century. Under this system, trees were felled with an ax sometime during the summer months, late enough in the season to discourage sprouting from stumps. They were allowed to lie until the following spring. Then, during the driest part of May, fire was put to the scattered heaps of wood and leaves. All but the trunks would burn, and the charred remnants of these could be sawed in half, piled together, and set afire once more. Such burning had several advantages over simple girdling. Although it destroyed much of the forest humus, it also killed the green roots of trees, preventing sprouting and enabling plowing to be done at an earlier date. With their roots dead, those stumps which were not turned to ashes rotted more quickly, and so could be removed sooner. Moreover, ashes returned much of the trees' nutrients to the soil, providing an effective fertilizer for the first year and substituting—at least in the short term—for any destruction of the humus layer. Unlike girdling, which returned nutrients to the soil over an extended period of time, burning did so in a single concentrated pulse, sacrificing longer-term conservation for shorter-term gain. Maize could be planted at once in the blackened field, without any need for plowing, hoeing, or manuring, and yielded a good crop in the first year. The second year, winter rye could be planted; thereafter, the field could be used for European grains, or, if seeded with grass, for pasture or mowing lands.[15]

Market conditions might change this procedure in important ways. If local demand were great enough, cut trees could be sold for lumber and a profit made from clearing itself. Where this was done, many of the fertilizing benefits of ashes were lost, and plowing had to be undertaken if the soil was to yield crops at all. Generally, only the well-to-do, who could purchase tools and labor, could afford such techniques, even though less timber was wasted when they were applied. If, on the other hand, a nearby market for potash or charcoal existed, even settlers who burned their lands could profit by selling ashes. Potash, which was used to manufacture soap and gunpowder, was a major New England export and in some regions furnished the sole cash crop during the initial year of settlement: it was claimed in 1717 that a farm laborer could, during the odd moments of a year, burn and process four acres of forest to produce eight tons of potash worth £40 to £60 per ton. Since the "improved" and newly fertilized land which resulted from such clearing could itself be sold for a profit, one understands why Jeremy Belknap could argue that it was "accounted more profitable for a young man to go upon new, than to remain on the old lands. In the early part of life, every day's labor employed in subduing the wilderness, lays a foundation for future profit." Destroying the forest thus became an end in itself, and clearing techniques designed to extract quick profits from forest resources encouraged movement onto new lands.[16]

The use of fire to aid in clearing land was something English settlers borrowed from their Indian predecessors, but they applied it for different purposes and on a much more extensive scale. Instead of burning the forest to remove undergrowth, they burned it to remove the forest itself. Doing so was not only profligate, consuming huge quantities of increasingly valuable timber, but dangerous as well. Within a year of settlement at Massachusetts Bay, John Winthrop reported that several haystacks and houses had been destroyed by the careless burning of fields. In an effort to avert such losses, the General Court in 1631 forbade colonists to fire ground before the first of March; Plymouth passed a similar law in 1633. Toward the end of the 1630s, both colonies established fire regulations limiting burning to specified weeks during the dampest spring months. Anyone damaging another person's property by uncontrolled burning was liable for the full extent of damages, and users of fire were required to warn their neighbors whenever they planned to burn their fields. These restrictions set the general pattern for similar laws in other New England colonies.[17]

Colonists thus modified the Indians' practice of large-scale communal burning in order to accommodate it to European notions of fixed property boundaries; fire was not to trespass across such boundaries under penalty of law. Inevitably, the Indians themselves were affected by the change. In 1640, for instance, the Narragansetts agreed to be liable for any harm done by their fires to colonial lands, the damages to be determined in an English court under English law. Such restrictions, which were one of several ways colonists taught Indians how Europeans bounded

the landscape, made earlier Indian uses of fire increasingly difficult to continue as colonial settlement advanced.[18]

New England towns required a regular supply of wood long
after their fields were cleared, and their efforts to obtain it extended the process of deforestation. Sawmills frequently became
the nuclei for new settlements in wooded areas, furnishing lumber for ships, churches, houses, barns, and all manner of farm
outbuildings. A town's roads and highways often converged on
its mills, which by cutting wood and grinding grain provided
vital economic services for the whole community. Although
many mills were initially simple sawpits in which two men cut
planks by hand, the high costs of skilled labor led a number of
New England mill operators to adopt water-power technology
well in advance of their English counterparts. Milldams created
artificial ponds, and mill activities cycled with the seasonal availability of water. Sawing charges varied depending on the species
of tree being cut—oak, for instance, cost twice as much as pine
—because the respective hardnesses of different woods required
that different amounts of labor and water be used in sawing
them. Even when mills ran from sunrise to sunset during seasons, like winter, when water was plentiful, their output probably averaged no more than a few hundred feet of lumber per day.
Such low productivity no doubt encouraged sawmills to use only
the best timber and waste the rest.[19]

In the face of initially abundant timber supplies, colonists altered many Old World uses of wood which had originally been
based on scarcity. Half-timbered construction of a building's
walls rapidly gave way to full-timbered construction using clapboards; stone-walled construction became relatively rare. Thatch
and slate roofs were replaced with wooden shingles. House size
in general increased over English models, so that buildings not
only required more lumber to build but more firewood to heat.
Even where bricks replaced lumber in construction, great quantities of wood were needed for firing their clay. In short, most
aspects of colonial house carpentry came to rely on the seemingly
endless supply of timber.[20]

Fences, which in England were usually made of stone or of
living hedges, in New England were initially made entirely of
wood. In part because they were designed to save as much labor
as possible, they too consumed large sections of the forest. The

first fences a farmer erected after clearing might simply consist of a row of stumps and large logs, or a worm fence of timbers stacked atop each other in a zigzag pattern. These were eventually replaced by rail or picket fences, which were used until repeated plowing turned up the rocks from which New England's famed stone walls were finally built. Most colonial wooden fences were poorly made, subject to rot, and wasteful of wood. As St. John de Crèvecœur remarked, "Our present modes of making fences are very bad . . . . They decay so fast, they are so subject to be hove up by the frost, it is inconceivable the cost and care which a large farm requires in that single article." To slow decay, colonists favored and selectively cut chestnut and cedar for fence construction, but frequently had to use oak. Where they did, a fence might last for no more than six to eight years before it had to be replaced. The final shift to stone walls was thus a way both of ending the labor cost of repeated fence construction and of conserving disappearing timber resources— not to mention getting rid of the annoying rocks which were accumulating along the edges of fields. "For the last few years," wrote one observer in Goshen, Connecticut, in the early nineteenth century, "there has been an increased attention to the building of stone fences; till which time chestnut rails were mostly used and the timber was fast decreasing."[21]

The greatest use of the New England forest by far, however, beyond fences or buildings or exported ship's masts, was for fuel. In addition to their love of large fires and warm houses, which had been apparent from the 1630s onward, New Englanders burned their wood in open fireplaces, which were four or five times less efficient than the closed cast-iron stoves of the Pennsylvania Germans. European travelers were frequently astonished by American consumption of firewood: the Swedish naturalist Peter Kalm remarked with some horror that "an incredible amount of wood is really squandered in this country for fuel; day and night all winter, or for nearly half of the year, in all rooms, a fire is kept going." A typical New England household probably consumed as much as thirty or forty cords of firewood per year, which can best be visualized as a stack of wood four feet wide, four feet high, and three hundred feet long; obtaining such a woodpile meant cutting more than an acre of forest each year. In 1800, the region burned perhaps eighteen times more wood for fuel than it cut for lumber. When the effects of such burning are summed up for the whole colonial period, it is probable that New England consumed more than 260 million cords of firewood between 1630 and 1800.[22]

This enormous demand could only be supplied by the forests. Farmers usually tried to maintain a woodlot on a hillside above their houses so that fuel could easily be dragged downhill for burning. Just as with lumber, certain species in such lots—especially hickory and oak—were preferred for firewood, so that these were depleted sooner than others. Be this as it may, not everyone had woodlots, and not all woodlots lasted forever. Local firewood scarcities often became a cause for concern within ten or fifteen years of a town's establishment. Boston, whose admittedly peculiar situation forced it from the start to gather fuel from the islands of Massachusetts Bay, experienced shortages as early as 1638, but other communities eventually did too. Private cutting of wood on common lands became a perennial source of dispute, and towns and colonies alike attempted to regulate it to prevent deforestation. In the long run, however, especially along the coast and near larger towns, timber for fuel became scarce and had to be obtained from greater and greater distances. Dwight wrote of one Vermont town whose inhabitants had "cut down their forest with an improvident hand: an evil but too common in most parts of this country. Unhappily it is an increasing evil, and may hereafter put a final stop to the progress of population." Elsewhere, he could report that in the 240-mile journey from Boston to New York a traveler passed through no more than twenty miles of wooded land, in fifty or sixty parcels. One result was a rise in fuel prices: the English traveler William Strickland was told in 1794 that timber and wood had "doubled in price, in every part of New England, within ten years." Most major cities of the East Coast shared this problem, which was a key reason for their eventual shift to coal in the nineteenth century.[23]

Timber scarcities and fuel shortages were of course local, as much a function of transportation costs as of absolute ecological changes. Great forests continued to exist in areas remote from rivers and population centers, but they were located at increasing distances from the regions undergoing deforestation. There, rising wood prices were a direct result of the prodigious consumption entailed by American clearing, lumbering, and heating practices. For observers accustomed to a world of scarcer resources, such uses of the forest seemed reckless in the extreme. Peter Kalm's judgment in 1749 retains much of its original force: "We can hardly be more hostile toward our woods in Sweden and Finland," he said, "than they are here: their eyes are fixed upon the present gain, and they are blind to the future."[24]

The ecological effects of this regional deforestation were profound, extending even to the climate itself. Although clearing changed little or nothing about the larger atmospheric movements of wind, clouds, or rain, it brought substantial changes at ground level in the way ecological communities experienced atmospheric phenomena. Microclimates, hydrology, and soil mechanics were all altered by the clearing process. It is difficult to quantify these changes with any precision, given the data that are available to us: as the cautious Dwight rightly pointed out, "Observation of this subject has been so loose, and the records are so few and imperfect, as to leave our real knowledge of it very limited." Broad trends nevertheless seem clear.[25]

Most New England naturalists agreed by the 1790s that deforestation and agricultural cultivation had the effect of warming and drying the soil, making the surface of the land hotter in summer and colder in winter. Temperatures in general fluctuated more widely without the moderating effects of the forest canopy to shade the ground and protect it from winds. Samuel Williams, in his 1809 *History of Vermont,* asserted that "the earth is no sooner laid open to the influence of the sun and winds, than the effects of cultivation begin to appear. The surface of the earth becomes more *warm* and *dry.*" He went so far as to demonstrate these effects with a series of experiments. Measuring soil temperatures in adjacent woodlots and pastures, he found that "the earth and the air, in the cultivated parts of the country, are heated in consequence of their cultivation, ten or eleven degrees more, than they were in their uncultivated state." He found in addition that a bowl of water placed in an open pasture evaporated one and a half times as quickly as one placed in the shade of a forest. More recent research has generally confirmed Williams's findings: forests tend to keep the ground beneath them cooler on average than open areas, and to narrow the range between maximum and minimum temperatures to produce a steadier climate.

They reduce wind speeds by 20 to 60 percent, and lower evaporation rates from soil. All these effects of course vary considerably, depending on the species and soil composition of the forest community, but all are interrupted by clearing. Cleared lands in colonial New England were thus sunnier, windier, hotter, colder, and drier than they had been in their former state.[26]

In wintertime, the effects of clearing produced an even more complex set of changes in these relationships. Although cleared land tended to be colder in winter than forested land—because drier and more exposed to the effects of wind chill—it received enough radiant heat from the sun to melt snow more quickly. The result was to shorten the apparent length of winter. As Dwight explained:

> In many cases, the first considerable snow will in a forested country become the commencement of winter; when, if the same country were generally open, the same snow would be wholly dissolved by the immediate action of the sun, and the winter in the appropriate sense would commence at a later period. On forested ground, also, the snow will lie later in the spring for the same reason. . . . Thus the summer half of the year must in such a country be somewhat shorter than if the forests were removed.

It was not, as some thought, that the weather itself was changed by clearing, but rather the way landscapes responded to the weather. If seasons were defined as much by an ecosystem's cycle of biological rhythms as by the movement of winds and storms, then in one special but important sense destroying the forest changed the very seasons themselves.[27]

Two major consequences followed when snow melted on cleared lands. For one, the longer retention of snow in forested areas acted to keep fires from spreading in the early months of spring; cleared lands were thus more susceptible to the burning colonists used to remove woody vegetation from them. More importantly in the long run, because the snow acted as a blanket maintaining soil temperatures, snow removal meant that soil froze to depths that it rarely had when forested. The frozen ground was unable to retain the water of melting snowfalls as easily as it once had; this effect was further compounded by the absence of tree roots and forest litter which had previously allowed soil to hold back the flow of large quantities of water. For all these reasons, spring runoff in deforested regions began and peaked at an earlier date; moreover, smaller rainstorms at other times of the year produced greater amounts of runoff. Watersheds emptied themselves more quickly, with the result that flooding became more common. Dwight described how the Connecticut River was "now often fuller than it probably ever was before the country above was cleared of its forests, the snows in open ground melting much more suddenly and forming much greater freshets than in forested ground." Negative evidence is always dangerous to use, but the most thorough history of New England storms records only one major flood in the region between 1635 and 1720; between 1720 and 1800, on the other hand, there were at least six that produced significant damage to life and property.[28]

Floods were a dramatic result of deforestation, but they were only the noisy heralds of a much subtler and more important change. In precolonial times, forests had not only held back the spring floodwaters with their roots and unfrozen ground; they had also staggered the runoff of that water over all the months of the year. Just as temperatures remained steadier in forested areas, so too did stream flows. When snowmelt and stormwater drained off cleared lands as floods, they left less water behind to keep streams and rivers running throughout the year. By the late eighteenth century, a number of naturalists were noticing how New England watersheds were responding to these changes. Noah Webster summarized them—using his own peculiar spelling system—as follows:

> The amazing difference in the state of a cultivated and uncultivated surface of erth, iz demonstrated by the number of small streems of water, which are dried up by cleering away forests. The quantity of water, falling upon the surface, may be the same; but when land iz cuvered with trees and leevs, it retains the water; when it iz cleered, the water runs off suddenly into the large streems. It iz for this reezon that freshes [floods] in rivers hav become larger, more frequent, sudden and destructiv, than they were formerly.

Floods, in other words, went hand in hand with dried-up streams and springs.[29]

Clearing also produced opposite results, ones that colonial observers rarely recorded. Forests lose an enormous amount of moisture through transpiration from their leaves; leaves also catch rainfall, which evaporates before it reaches the ground. Water lost in these ways can make up a large share of the annual precipitation that falls on a forested area. Removing trees thus actually *increases* the total amount of water flowing off the land into streams and rivers. In low, poorly drained areas, colonial clearing sometimes had the effect of transforming a relatively dry area into a swamp. This may be one explanation for the widespread tendency among those who visited frontier settlements to link the process of clearing with disease. As Dwight observed, "While the country is entirely forested, it is ordinarily healthy. While it is passing from this state into that of general cultivation, it is usually less healthy." Colonists attributed their "fevers" and "agues" to bad air and miasma rising from newly exposed soil; in fact, the real culprit may have been anopheles mosquitoes carrying malaria, their populations temporarily swelled by newly swampy areas which had not been drained.[30]

In the long run, however, even though more water entered drainage systems as a result of deforestation, its irregular and more rapid runoff left the countryside drier at most seasons of the year than it had been before. "One of the first effects of cultivation," said Ira Allen in his *History of Vermont*, "is the dissipation of the waters." The drying up of streams and springs continued for decades after their forests were removed, and their eventual disappearance could mean economic crisis for the farms and towns which had depended on them. "In many parts of those States which have been cleared for above a generation or two," wrote the Vermont naturalist George Perkins Marsh in 1864, "the hill pastures now suffer severely from drought, and in dry seasons no longer afford either water or herbage for cattle." Peter Whitney mentioned a pond in Worcester, Massachusetts, whose size had been cut in half by more irregular drainage patterns, and there were undoubtedly other such cases elsewhere. Probably the colonial enterprise that suffered most from these changes was the one which relied most heavily on waterpower. Mills, which had depended on small streams and ponds to turn their wheels, frequently found themselves first with not enough water to work in summer, and finally with little water at all save in periods of peak runoff—the very moment that they were threatened with damage by floods. As Samuel Williams described the phenomenon, "Mills, which at the first settlement of the country, were plentifully supplied with water from small rivers, have ceased to be useful."[31]

In summary, then, deforestation was one of the most sweeping transformations wrought by European settlement in New England. It aided in the reduction of edge-dwelling animal species. Where forests were not completely eliminated, their species composition changed: trees such as white pines, white cedars, and white oaks became less common. Where forests were entirely destroyed, the landscape became hotter in summer and colder in winter. Temperatures in general fluctuated more widely. Snow melted more readily than it had before, and the ground froze more deeply. The water-holding capacity of the soil was reduced, and runoff was thereby increased and made more erratic. Flooding became more common, and stream levels came to vary so greatly that some dried up altogether for extended seasons of the year. Water tables fell. But the list does not stop even here.[32]

Dramatic as these changes may be, their full effect remains invisible until they can be seen in relation to the ecological habitats with which Europeans replaced the vanished forests. It was no accident that the colonists cleared land so much more extensively than the Indians had done, nor was it mere chance that the English had such destructive effects on New England forests. The colonists themselves understood what they were doing almost wholly in positive terms, not as "deforestation," but as "the progress of cultivation." The two descriptions were in reality simply inverse ways of stating a single fact: the rural economies of Europe were adapted to a far different mosaic of ecological habitats than were precolonial Indian economies. Reducing the forest was an essential first step toward reproducing that Old World mosaic in an American environment. For the New England landscape, and for the Indians, what followed was undoubtedly a new ecological order; for the colonists, on the other hand, it was an old and familiar way of life.

7

A WORLD OF
FIELDS AND FENCES

One must not exaggerate the differences between English and Indian agricultures. The two in many ways resembled each other in the annual cycles by which they tracked the seasons of the year. Although English fields, unlike Indian ones, were cultivated by men as well as women and contained a variety of European grains and garden plants which were segregated into single-species plots, their most important crop was the same maize grown by Indians. Like the Indians, the English began working their fields when the land thawed and cleared of snow sometime in March. They too planted in late March, April, and May, and weeded and hilled their corn—if rather more carelessly than the Indians—a month or two later. Summer saw colonists as well as Indians turning to a wide range of different food sources as they became available: fish, shellfish, migratory birds, foraging mammals, and New England's many wild berries. August through October was the season of harvest when corn was gathered, husked, and stored, and other crops were made ready for the winter months. November and December saw the killing of large mammals—albeit of different species than the Indians had hunted—from the New England woods, the meat and hides of which were then processed for use in the months to come. The rest of the winter was devoted to tasks any southern New England Indian would readily have recognized: making and repairing tools and clothing, looking after firewood, occasionally fishing or hunting, and generally living off the stored produce of the preceding year. As the days lengthened and became warmer, the cycle began again: Europeans as well as Indians were inextricably bound to the wheel of the seasons.[1]

What made Indian and European subsistence cycles seem so different from one another had less to do with their use of plants than their use of animals. Domesticated grazing mammals—and the tool which they made possible, the plow—were arguably the single most distinguishing characteristic of European agricultural practices. The Indians' relationships to the deer, moose, and beaver they hunted were far different from those of the Europeans to the pigs, cows, sheep, and horses they owned. Where Indians had contented themselves with burning the woods and concentrating their hunting in the fall and winter months, the English sought a much more total and year-round control over their animals' lives. The effects of that control ramified through most aspects of New England's rural economy, and by the end of the colonial period were responsible for a host of changes in the New England landscape: the seemingly endless miles of fences, the silenced voices of vanished wolves, the system of country roads, and the new fields filled with clover, grass, and buttercups.[2]

Livestock were initially so rare in Plymouth and Massachusetts Bay that both William Bradford and John Winthrop noted in their journals the arrival of each new shipment of animals. Plymouth was over three years old before it obtained the "three heifers and a bull" which Bradford described as "the first beginning of any cattle of that kind in the land." Massachusetts Bay had a larger number of livestock almost from the start, but there were by no means enough to satisfy colonial demand for more animals. One colonist explained to an English patron that cattle were "wonderful dear here," and another argued that the most profitable investment a merchant could make in New England would be "to venture a sum of moneys to be turned into cattle." As a result, ship after ship arrived laden with upward of fifty animals in a load. By 1634, William Wood was able to define the wealth of the Massachusetts Bay Colony simply by referring to its livestock. "Can they be very poor," he asked, "where for four thousand souls there are fifteen hundred head of cattle, besides four thousand goats and swine innumerable?"[3]

The importance of these various animals to the English colonists can hardly be exaggerated. Hogs had the great virtues of reproducing themselves in large numbers and—like goats of being willing to eat virtually anything. Moreover, in contrast to most other English animals, they were generally able to hold their own against wolves and bears, so that they could be turned out into the woods for months at a time to fend for themselves almost as wild animals. They required almost no attention until the fall slaughter, when—much as deer had been hunted by Indians—they could be recaptured, butchered, and used for winter meat supplies. Cattle needed somewhat more attention, but they too were allowed to graze freely during the warmer months of the year. In addition to the meat which they furnished, their hides were a principal source of leather, and milch cows provided dairy products—milk, cheese, and butter—that were unknown to the Indians. Perhaps most importantly, oxen were a source of animal power for plowing, clearing, and other farmwork. The use of such animals ultimately enabled English farmers to till much larger acreages than Indians had done, and so produce greater marketable surpluses. When oxen were attached to wheeled vehicles, those surpluses could be taken to market and sold. Horses were another, speedier source of power, but they were at first not as numerous as oxen because they were used less for farmwork than for personal transportation and military purposes. Finally, sheep, which required special attention because of their heavy wear on pastures and their vulnerability to predators, were the crucial supplier of the wool which furnished (with flax) most colonial clothing. Each of these animals in its own way represented a significant departure from Indian subsistence practices.[4]

What most distinguished a hog or a cow from the deer hunted by Indians was the fact that the colonists' animal was owned. Even when it grazed in a common herd or wandered loose in woodlands or open pastures, a fixed property right inhered in it. The notch in its ear or the brand on its flanks signified to the colonists that no one other than its owner had the right to kill or convey rights to it. Since Indian property systems granted rights of personal ownership to an animal only at the moment it was killed, there was naturally some initial conflict between the two legal systems concerning the new beasts brought by the English. In 1631, for instance, colonists complained to the sachem Chickatabot that one of his villagers had shot an English pig. After a month of investigation, a colonial court ordered that a fine of one beaver skin be paid for the animal. Although the fine was paid by Chickatabot rather than the actual offender—suggesting the confusion between diplomatic relations and legal claims which necessarily accompanied any dispute between Indian and English communities—the effect of his action was to acknowledge the English right to own animal flesh. Connecticut went so far as to declare that Indian villages adjacent to English ones would be held liable for "such trespasses as shalbe committed by any Indian"—whether a member of the village or not—"either by spoilinge or killinge of Cattle or Swine either with Trappes, dogges or arrowes." Despite such statutes, colonists continued for many years to complain that Indians were stealing their stock. As late as 1672, the Massachusetts Court was noting that Indians "doe frequently sell porke to the English, and there is ground to suspect that some of the Indians doe steale and sell the English mens swine." Nevertheless, most Indians appear to have recognized fairly quickly the colonists' legal right to own animals.[5]

Ironically, legal disputes over livestock arose as frequently when Indians *acknowledged* English property rights as when they denied them. One inevitable consequence of an English agricultural system that mixed the raising of crops with the keeping of animals was the necessity of separating the two—or else the animals would eat the crops. The obvious means for accomplishing this task was the fence, which to colonists represented perhaps the most visible symbol of an "improved" landscape: when John Winthrop had denied that Indians possessed anything more than a "natural" right to property in New England, he had done so by arguing that "they inclose noe Land" and had no "tame Cattle to improve the Land by." Fences and livestock were thus pivotal elements in the English rationale for taking Indian lands. But this rationale could cut both ways. If the absence of "improvements"—fences—meant to the colonists that Indians could claim to own only their cornfields, it also meant that those same cornfields lay open to the ravages of English grazing animals. Indians were quick to point out that, since colonists claimed ownership of the animals, colonists should be responsible for any and all damages caused by them.[6]

Much as they might have preferred not to, the English had to admit the justice of this argument, which after all followed unavoidably from English conceptions of animal property. Colonial courts repeatedly sought some mechanism for resolving the perennial conflict between English grazing animals and Indian planting fields. In 1634, for instance, the Massachusetts Court sent an investigator "to examine what hurt the swyne of Charlton hath done amongst the Indean barnes of corne," and declared that "accordingly the inhabitants of Charlton promiseth to give them satisfaction." Courts regularly ordered payment of compensation to Indians whose crops had been damaged by stock, but this was necessarily a temporary solution, administered after the fact, and one which did nothing to prevent further incidents. Colonists for this reason sometimes found themselves building fences on behalf of Indian villages: in 1653, the town of New Haven promised to contribute sixty days of labor toward the construction of fences around fields planted by neighboring Indians. Similar efforts were undertaken by colonists in Plymouth Colony when Indians at Rehoboth complained of the "great dammage" caused to their crops by English horses. The fences built across the Indians' peninsula of land at Rehoboth did not, of course, prevent animals from swimming around the barrier, and so Plymouth eventually—for a short while—granted Indians the right to impound English livestock and demand payment of damages and a fine before animals were returned to their owners.[7]

Such solutions had the virtue—in principle—of giving Indians at least theoretical legal standing as they made their complaints to colonial courts, but the laws also forced Indians to conduct their agriculture in a new way. Indians wishing compensation for damages to their crops were required to capture wandering animals and hold them until they were claimed by their owners; moreover, the value of damages was to be "judged and levied by some indifferent man of the English, chosen by the Indians treas-

pased." Indians had no right to collect damages by killing a
trespassing animal: as always, the rare occasions when colonists
offered Indians legal protection were governed wholly by English terms. The long-run effect was to force Indians to adopt
fencing as a farming strategy. When New Haven built fences for
its neighbors, it was only on the understanding that the Indians
would agree "to doe no damag to the English Cattell, and to
secure their owne corne from damage or to require none." Indians, in other words, were eventually assumed to be liable for the
maintenance of their own fences; if these were in disrepair, no
damages could be collected for the intrusions of wandering animals. When Indians near Massachusetts Bay "promised against
the nexte yeare, and soe ever after, to fence their corne against
all kinde of cattell," they were making what would prove to be
an irrevocable commitment to a new way of life.[8]

Indians were not alone among New England's original inhabitants in encountering new boundaries and conflicts as a result of
the colonists' grazing animals. Native predators—especially
wolves—naturally regarded livestock as potential prey which
differed from the deer on which they had previously fed only by
being easier to kill. It is not unlikely that wolves became more
numerous as a result of the new sources of food colonists had
inadvertently made available to them—with unhappy consequences for English herds. Few things irritated colonists more
than finding valuable animals killed by "such ravenous cruell
creatures." The Massachusetts Court in 1645 complained of "the
great losse and damage" suffered by the colony because wolves
killed "so great nombers of our cattle," and expressed frustration
that the predators had not yet been successfully destroyed. Such
complaints persisted in newly settled areas throughout the colonial period.[9]

Colonists countered the wolf threat in a variety of ways. Most
common was to offer a bounty—sometimes twopence, sometimes
ten shillings, sometimes a few bushels of corn, sometimes (for
Indian wolf hunters) an allotment of gunpowder and shot—to
anyone who brought in the head of a wolf. All livestock owners
in a town were legally bound to contribute to these bounties.
Their effect, like that of the fur trade, was to establish a price for
wild animals, to create a court-ordered market for them, and so
encourage their destruction. Indians especially were led to hunt

wolves by these means. But bounties had several drawbacks. They occcasionally tempted hunters to bring in the heads of wolves which had been killed many miles from English settlements and so forced towns to pay for predators that had never been a threat to English stock. Since neither wolves nor Indians respected the jurisdictional boundaries of English towns, it was difficult to distinguish which town should pay which hunter for which wolf. There was, moreover, a recurring problem of people trying to sell the same wolf's head twice, so that towns were forced to cut off the dead animal's ears and bury them separately from the skull in order to prevent repeated bounty payments for it.[10]

Whenever the wolf problem seemed especially severe, colonists supplemented the customary bounties with more drastic measures. Special hunters were occasionally appointed to leave poisoned bait or to set guns with tripwires as traps for wolves; domesticated animals killed by the traps were to be paid for by the town as a whole. Individual wolves that were particularly rapacious might have an unusually high bounty placed on their heads: thus, New Haven in 1657 offered five pounds to anyone who could kill "one great black woolfe of a more than ordinarie bigness, which is like to be more feirce and bould than the rest, and so occasions the more hurt." A number of settlements fought wild dogs with domesticated ones. In 1648, Massachusetts ordered its towns to procure "so many hounds as they thinke meete . . . that so all meanes may be improved for the destruction of wolves." Towns regularly ordered collective hunts of areas that were suspected of harboring wolves. Indeed, wolves became a justification for draining and clearing swamps. The anonymous "Essay on the Ordering of Towns" in 1635 suggested that towns hold every inhabitant responsible for clearing the "harboring stuffe" from the "Swampes and such Rubbish waest grownds" that sheltered wolves. That this was not merely a theoretical proposal is shown by Scituate's decision in the mid-seventeenth century to divide its five hundred acres of swampland into parcels of two to five acres and to require every landowner to clear one of them. Entire ecological communities were thus threatened because they represented "an annoyance and prejudice to the town . . . both by miring of cattle and sheltering of wolves and vermin." Whether the assault was conducted with bounties, hunting dogs, or the removal of the animals' habitats, wolves suffered much the same fate as the rest of New England's mammals. Although they were still found in northern New England at the end of the colonial period, Dwight reported that they were gone from the south. Because, unlike Indians, wolves were incapable of distinguishing an owned animal from a wild one, the drawing of new property boundaries on the New England landscape inevitably meant their death.[11]

Conflicts among Indians, wolves, and colonists over the possession and trespasses of grazing animals paralleled even more complicated conflicts among colonists themselves. Farmers accustomed to an English landscape hemmed in by hedges and by customs regulating the common grazing of herds found themselves forced to change their agricultural practices in unfenced New England. Whereas most livestock in England had been watched over by individual herders, labor was scarce enough in New England that only the most valuable animals—milch cows, sheep, and some horses and oxen—could generally be guarded in this way. Large numbers of swine and dry cattle in particular received less supervision than they would have gotten in England, and so presented a more or less constant threat to croplands. Fences could take the place of herders only where colonists built and maintained them, conditions that rarely applied in new settlements. Accordingly, towns and colonies alike were constantly shifting their regulations in an effort to control the relationship between domesticated animals and crops.

Cattle and horses, for instance, were valuable enough so that colonial laws usually held farmers responsible for protecting their crops from them rather than requiring that the animals be restrained. In 1633, the Plymouth Court ordered that no one should "set corne . . . without inclosure but at his perill." Massachusetts Bay tried in 1638 to spread the burden of this responsibility by requiring that "they that plant are to secure theire corne in the day time; but if the cattle do hurt corne in the night, the owners of the cattle shall make good the damages." But the new rule, not too surprisingly, proved impossible to enforce, and within two years the colony was again declaring that planters rather than animal owners must bear the responsibility for protecting crops. Unfenced property boundaries, in other words, did not give legal protection against trespass by cattle. According to the Massachusetts Court in 1642, "Every man must secure his corne and medowe against great cattell." If a property owner failed to do this, said the Court, and "if any damage bee done by such cattell, it shallbe borne by him through whose insufficient fence the cattle did enter."[12]

But this by no means solved all problems. Large animals were quite capable of destroying a sound fence if they put their minds to it, and so towns were eventually forced to appoint fence viewers, who regularly visited farms to "see that the fence be sett in good repaire, or else complaine of it." If a fence was declared sound by the fence viewers, its owner could collect damages from anyone whose animal broke into the field; if a fence was unsound, on the other hand, its owner not only was legally unprotected from damage by animals but might also be required to pay the costs borne by neighbors who repaired it. The competing claims of property in animals and property in lands were thus resolved in a way that in principle seemed to reduce the absolute protection of the latter. In practice, however, cattle regulations had the opposite effect. Through the agency of the fence viewers and the formal litigation of the courts, towns took an increasing responsibility not only for enforcing the abstract boundaries between adjacent tracts of real estate but for guaranteeing that those boundaries were marked by the physical presence of fences. The fact that landed property received only conditional protection in law became a major impetus for fencing the countryside, and so redrawing the New England map.[13]

Not all English animals were equally protected by such fence laws. Swine were the weed creatures of New England, breeding so quickly that a sow might farrow three times in a year, with each litter containing twelve to sixteen piglets. They so rapidly became a nuisance that, as early as 1633, the Massachusetts Court declared that "it shalbe lawfull for any man to kill any swine that comes into his corne"; the dead animal was to be returned to its owner only after payment had been made for damages to crops. The inadequacy of this solution is suggested by the proliferation of swine laws in the ensuing years. Colonists were glad to have swine reproducing and fattening themselves in forested areas distant from English settlements—where only Indians would have to deal with their depredations—but towns tried to restrain the animals whenever they wandered too near English fields. By

1635, Massachusetts had ordered towns to construct animal pounds to which untended swine could be taken whenever they were found within one mile of an English farm. A year later, the Court went so far as to declare an open season on any stray swine: unless pigs were restrained by fence, line, or pigkeeper, it was lawful "for any man to take them, either alive or dead, as hee may." Anyone so doing got one half the value of the captured animal, while the Commonwealth of Massachusetts claimed the other; the owner got nothing. Ownership rights to swine were thus much more circumscribed than similar rights to cattle. The law produced so much protest from pig-keeping colonists that it was repealed two years later, but the battle of the swine nevertheless continued for many years. Complaints against pigs were a near constant feature of colony and town court proceedings, where the animals were sometimes portrayed almost as a malevolent force laying siege to defenseless settlements. The Massachusetts Court in 1658, for instance, reported that "many children are exposed to great daingers of losse of life or limbe through the ravenousnese of swyne, and elder persons to no smale inconveniencies." To modern ears, such statements perhaps seem a little comic, but that reaction is surely one of ignorance: swine could indeed be vicious creatures, and no animal caused more annoyances or disputes among colonists.[14]

Ultimately, swine were relegated either to farmyard sties, where they could be fed corn, alewives, and garbage, or to relatively isolated areas, where they could feed as they wished and do little harm. Favorite swine-raising locations were coastal peninsulas and offshore islands, where the animals were free to do their worst without interfering with English crops. In the late 1630s, both Roger Williams and John Winthrop were moving swine onto islands in Narragansett Bay, and colonists elsewhere did likewise. Along the coast, the animals wreaked havoc with oyster banks and other Indian shellfish-gathering sites, but caused little trouble to the English. Roger Williams described how "the English swine dig and root these Clams wheresoever they come, and watch the low water (as the Indian women do)." In one important sense, then, English pigs came into direct competition with Indians for food: according to Williams, "Of all English Cattell, the Swine (as also because of their filthy dispositions) are most hateful to all Natives, and they call them filthy cut throats." Pigs thus became both the agents and the emblems for a European colonialism that was systematically reorganizing Indian ecological relationships.[15]

In the vicinity of English settlements, regulations were eventually passed requiring that hogs be yoked so that they would be unable to squeeze through fences, and ringed through the nose so that they would be prevented from rooting out growing plants. But the chief goal of swine regulations was to keep uncontrolled pigs away from settlements. At a heated New Haven town meeting in 1650, farmers declared that, if swine were allowed to forage freely, "they would plant no corne, for it would be eaten up." The compromise solution was an order that no pigs should run loose unless driven at least eight miles from town center. Other communities passed similar regulations. And yet driving swine to the edges of town was obviously a temporary solution that lasted only so long as a town had edges beyond which were unenclosed common lands where pigs could run. Moreover, this "solution" tended to provoke conflict *between* towns when swine crossed town boundaries to descend on other settlements. Massachusetts Bay in 1637 pointed to the long-term solution of this problem by disclaiming any direct responsibility for the regulation of swine and delegating that burden to individual towns. "If any damage bee done by any swine," it said, "the whole towne shalbee lyable to the parties action to make full satisfaction." By making the control of swine a community responsibility, the Court redefined the property boundaries that applied to this particular animal so as to ensure its proper regulation. As the landscape gradually became peopled with settlements, the effect of legal liabilities was increasingly to restrain the movements of wandering hogs, until finally the beasts were more or less entirely confined to fenced farmyards.[16]

What became true of swine also became true of horses, sheep, and cattle: each was allocated its separate section of a settlement's lands. The interactions among domesticated grazing animals, demographic expansion, and English property systems had the effect not only of bounding the land with relatively permanent fences but of segregating the uses to which that land was put. Even the earliest colonial towns had divided their territories according to intended function, and colonists had been granted land accordingly. Fences thus marked off, not only the map of a settlement's property rights, but its economic activities and ecological relationships as well. At the center of a family's holdings was its house lot, around which a host of activities revolved, most of them controlled by women: food processing, cloth and tool making, poultry keeping, vegetable and herb gardening, and domestic living generally. Nearby were the outbuildings where animals spent their winters and some of their summer nights, as well as the various lots in which sheep, horses, milch cows, and pigs could be fed when not free to graze. In order for such animals to survive the winter, hay had to be cut in mid- to late summer, dried, and rationed out to them from November through early spring. This necessitated reserving large tracts of land for mowing, an activity which generally took place along the banks of streams, in salt marshes, and anywhere else that grass could be found. Aside from grain fields, all other lands were committed to grazing, including the upland woodlots where families cut their fuel and lumber. The key functional boundary in an English settlement was always the one between pasture and nonpasture: it was because the barrier between these two had to be so rigid that colonial towns presented such a different appearance from that of earlier Indian villages.[17]

English colonists reproduced these broad categories of land use wherever and however they established farms. Early land divisions had been done communally, each town deciding what agricultural activity would take place in different parts of its territory. Later divisions were generally made through the abstract mechanism of land speculation and tended to ignore both the ecological characteristics of a given tract of land and its intended agricultural use in order to facilitate the buying and selling that brought profits to speculators. This marked an important new way of perceiving the New England landscape, one that turned land itself into a commodity, but from the point of view of ecological practices, it merely transferred land-use decisions from the town to the individual landowner. Every farm family had to have its garden, its cornfields, its meadows, and its pastures, no matter who decided where they would be located and how they would be regulated. In so dividing their lands, colonists began to create the new ecological mosaic that would gradually transform New England ecosystems.[18]

Livestock not only defined many of the boundaries colonists

drew but provided one of the chief reasons for extending those boundaries onto new lands. Indian villages had depended for much of their meat and clothing on wild foraging mammals such as deer and moose, animals whose populations were much less concentrated than their domesticated successors. Because there had been fewer of them in a given amount of territory, they had required less food and had had a smaller ecological effect on the land that fed them. The livestock of the colonists, on the other hand, required more land than all other agricultural activities put together. In a typical town, the land allocated to them was from two to ten times greater than that used for tillage. As their numbers increased—something that happened quite quickly—the animals came to exert pressure even on these large amounts of land.[19]

Before examining the ecological relationships of domesticated animals, it is well to remember their economic relationships. Livestock very early came to play a role in the New England economy comparable to that of fish and lumber: they proved to be a most reliable commodity. By 1660, Samuel Maverick, who had been one of the earliest English settlers in Massachusetts Bay, could point to increased numbers of grazing animals as one of the most significant changes in New England towns since his arrival. "In the yeare 1626 or thereabouts," he said,

> there was not a Neat Beast Horse or sheepe in the Countrey and a very few Goats or hoggs, and now it is a wonder to see the great herds of Catle belonging to every Towne. ... The brave Flocks of sheepe, The great number of Horses besides those many sent to Barbados and the other Carribe Islands, And withall to consider how many thousand Neate Beasts and Hoggs are yearly killed, and soe have been for many yeares past for Provision in the Countrey and sent abroad to supply Newfoundland, Barbados, Jamaica, and other places, As also to victuall in whole or in part most shipes which comes here.

Maverick viewed New England with a merchant's eye, and regarded its livestock as one of its most profitable productions.[20]

Whether sold fresh to urban markets or salted for shipment to Caribbean sugar plantations, grazing animals were one of the easiest ways for a colonist to obtain hard cash with a minimum of labor. October and November saw many colonial farmers make an annual pilgrimage to coastal cities such as Boston, New Haven, and Providence, where fatted animals could be sold or exchanged for manufactured goods. This economic profitability contributed to the ecological consequences of livestock raising. Besides intensifying pressure on grazing lands and inviting more territorial expansion, it necessitated the construction of roads connecting interior towns with urban centers. No small number of trees were destroyed by the construction of these roads—they were typically between 99 and 165 feet wide—but their seemingly excessive size was more than justified since they facilitated moving large herds to market. Roads were the link binding city and countryside into a single economy. During the course of the colonial period, the opportunities represented by that linkage encouraged farmers to orient more and more of their production toward commercial ends. As one eighteenth-century visitor to New England observed:

> Boston and the shipping are a market which enriches the country interest far more than the [trade in exports,] which, for so numerous a people, is very inconsiderable. By means of this internal circulation, the farmers and country gentlemen are enabled very amply to purchase whatever they want from abroad.

Almost from the start, port cities exemplified the different ways in which Indians and colonists organized their economies, and no commodity moved more readily from farm to city than did animals.[21]

Livestock production was tied to the markets of the ports by a web of relationships that extended well beyond the fall drives. Whether generating a surplus by their own reproduction, by their labor in working crops, or by their contribution to lowering transportation costs in bringing themselves and other goods to market, grazing animals were one of the linchpins that made commercial agriculture possible in New England. Without them, colonial surpluses would probably have been produced on much the same scale as Indian ones; with them, colonial agriculture had a more or less constant tendency to expand and to put increasing pressure on its surrounding environment. As the ecologist E. Fraser Darling has noted, "Pastoralism for commercial ends . . . cannot continue without progressive deterioration of the habitat."[22]

Signs that such deterioration was taking place, or at least that the number of animals was outrunning the available food supply, became apparent within four years of Boston's founding. In 1634, the inhabitants of Newtown (Cambridge) complained of "want of accommodation for their cattle," and asked the Massachusetts Court for permission to migrate to Connecticut. When colonists in Watertown and Roxbury put forward similar petitions a year later, John Winthrop explained that "the occasion of their desire to remove was, for that all towns in the bay began to be much straitened by their own nearness to one another, and their cattle being so much increased." Regions which had once supported Indian populations considerably larger than those of the early English settlements came to seem inadequate less because of *human* crowding than because of *animal* crowding. Competition for grazing lands—which were initially scarcer than they later became—acted as a centrifugal force that drove towns and settlements apart. In 1631, Bradford lamented the changes wrought by livestock in Plymouth Colony: "no man," he wrote,

> now thought he could live except he had cattle and a great deal of ground to keep them, all striving to increase their stocks. By which means they were scattered all over the Bay quickly and the town in which they lived compactly till now was left very thin and in a short time almost desolate.

Unlike tillage, whose land requirements were far lower, pastoralism became a significant force for expansion. Further, if Bradford is to be believed, it also contributed to the famous declension which helped drive New England towns from their original vision of compact settlements, communal orders, and cities upon hilltops.[23]

One reason that scarcity of grazing land so quickly became a problem in Massachusetts Bay had to do with the nature of New England's native grasses, which included broomstraw, wild rye, and the *Spartinas* of the salt marshes. Because most of the first English settlements were made on Indian village sites, the lands of which had been regularly cultivated and burned, there were extensive areas around them where only grass and shrubs grew. Animals could be turned loose to graze on these with virtually no preparation of the land, but often seemed to fare poorly on their new diet. Many colonists commented on the relative inferiority of New England hay in comparison with that of England, and one wrote in disgust that "it is so devoid of nutritive vertue, that our beasts grow lousy with feeding upon it, and are much out of heart and liking." More serious than the *quality* of the native grasses, however, was their inadequate *quantity:* domesticated animal populations quickly ran out of pasture, so that their owners had to clear land to create more.[24]

Curiously, many colonists claimed that the native grasses, although initially very "rank" and "coarse," seemed to improve the more they were mowed or eaten. "In such places where the cattle use to graze," wrote William Wood, "the ground is much improved in the woods, growing more grassy and less weedy." What in fact was happening was that a number of native grasses and field plants were slowly being destroyed and replaced by European species. Annual grasses were quickly killed off if grazed too closely, and the delicate crowns of some perennials fared little better. Not having evolved in a pastoral setting, they were ill prepared for their new use. That was why European grasses, which *had* adapted themselves to the harsh requirements of pastoralism, began to take over wherever cattle grazed. "English grasses," such as bluegrass and white clover, spread rapidly in newly settled areas. Initially carried to the New World in shipboard fodder, and in the dung of the animals which ate them, these European species were soon being systematically cultivated by colonists. By the 1640s, a regular market in grass seed existed in the Narragansett country, and within one or two generations, the plants had become so common that they were regarded as native.[25]

Grazing animals were among the chief agents in transmitting to America one of the central—albeit unapplauded—characters of European agriculture: the weed. Because Indians kept no cattle, and because their mixed-crop, hoe agriculture provided a relatively dense ground cover, they failed to develop as many of the plant species which in the Old World followed wherever human beings disturbed the soil. Like the "English grasses," weeds had evolved any of a number of adaptations that allowed them to tolerate grazing and to move quickly onto cleared agricultural land: they were able to germinate under a wide variety of environmental conditions, they grew rapidly, they might continuously produce huge quantities of seeds designed for widespread dispersal, and they were often brittle so that when broken off by cattle or farmers they could readily regenerate themselves from their remaining fragments. A few indigenous species had enough of these characteristics that they too became more common as a result of European settlement. Probably the most prolific of these was ragweed, which underwent such a population explosion in the colonial period that pollen scientists today, when studying the sediments in pond and lake bottoms, use the plant as a means of dating the arrival of the Europeans.[26]

Most weeds, however, were European. John Josselyn in 1672 listed no fewer than twenty-two European species which had become common in the area around Massachusetts Bay "since the English planted and kept Cattle in New England." Among these were such perennial favorites as dandelions, chickweeds, bloodworts, mulleins, mallows, nightshades, and stinging nettles. Because it seemed to crop up wherever the English walked, planted, or grazed animals, the Indians called plantain "Englishman's Foot," a name that suggests their awareness of the biological invasion going on around them. Not only Indians were affected by this invasion, since colonial grain crops—and, worse, the seeds used to plant them—were difficult to keep separate from the weeds that grew in their midst. As early as 1652, settlers in New Haven Colony were debating whether something could be done "to prevent the spreding of sorrill in the corne feilds," but did so to no avail. Many of these European weeds—to say nothing of grains, vegetables, and orchard trees—would eventually be among the commonest plants of the American landscape, their populations sustained in all places by the habitats human beings and domesticated animals created for them.[27]

Although the invasion of livestock was sustained by the parallel invasion of edible plants, the two were rarely in perfect balance, at least in the eyes of colonists who for economic reasons sought to raise more animals. Livestock production expanded throughout New England in the eighteenth century and brought with it regular complaints about pasture shortages. By 1748, the

Connecticut agricultural writer Jared Eliot was commenting that "the scarcity and high price of hay and corn is so obvious, that there are few or none Ignorant of it." The shortage of hay, he said, had been "gradually increasing upon us for sundry Years past," and was the direct result of livestock populations outgrowing available meadowlands. If pastures were inadequate, old and new settlements alike had to follow the process of forest clearing described in the preceding chapter, planting corn and rye before the unplowed soil was finally ready to be seeded with English grasses. During the eighteenth century, the range of grasses which were raised for mowing was extended to include such species as timothy, red clover, lucerne (alfalfa), and fowl-meadow grass, all of which rapidly became common throughout the colonies.[28]

Where mowing was unnecessary and grazing among living trees was possible, settlers saved labor by simply burning the forest undergrowth—much as the Indians had once done—and turning loose their cattle. But because English livestock grazed more closely and were kept in denser concentrations than the animals for whom Indians had burned the woods, English pastoralism had the effect of gradually shifting the species composition of any forest used for pasture. In at least one ill-favored area, the inhabitants of neighboring towns burned so frequently and grazed so intensively that, according to Peter Whitney, the timber "was greatly injured, and the land became hard to subdue. Hurtleberry and whitebush sprung up, together with laurel, sweetfern and checkerberry, which nothing but the plough will destroy." In the long run, cattle tended to encourage the growth of woody, thorn-bearing plants which they could not eat, and which, once established, were very difficult to remove. Such plants had to be cleared regularly with a scythe or grub hoe if they were not to take over a pasture entirely. The only other way of dealing with them was to graze sheep heavily in areas which the bushes had taken over; the flocks sometimes succeeded in reclaiming land that had otherwise become useless.[29]

The tree species of the uplands were also affected by grazing, especially when exhausted fields were allowed to revert to wooded pastures. Hemlocks, whose shallow root systems were very sensitive to fire, tended to disappear from all woods that were burned for pasture. Where they were protected from fire, on the other hand, grazing encouraged their growth by destroying the more edible hardwood species that would otherwise have competed with them, so that hemlocks then became the dominant species of north-facing slopes. When land was initially cleared, whether for crops or pasture, the removal of existing trees had the effect of releasing the dormant seeds of certain species that preferred full sunlight and open growing conditions. Pin cherry was one of these. Timothy Dwight told of a farmer in Vermont in whose fields "there customarily sprung up . . . an immense multitude of cherry trees," even though the surrounding forest was composed entirely of beech, hemlock, and maple. Red cedar also often acted as a pioneer on cleared lands.[30]

Which species invaded which fields depended primarily upon whether or not grazing animals were allowed on the land. The ecological effects of pasturing and clearing on forest composition could become quite complex. In oak and birch forests that were cut for lumber and fuel, for instance, these two tree species were able to regenerate themselves by sprouting from their roots and stumps, and could be cut again in as little as fourteen years. Cyclical cutting of this kind—known as a coppice system—was common among colonial farmers, and strongly favored hardwood species, which could sprout, over conifers, which could not. Coppice cutting was a major reason that chestnuts, which were prolific sprouters, increased their relative share of New England forests following European settlement. But if sprout hardwood forests were used for pasture after being cut, the sprouts were destroyed by being grazed, and the less edible white pine often came up instead. Conversely, white pines—which could not sprout but compensated for this by producing enormous quantities of airborne seeds—failed to regenerate themselves *unless* pasturing took place, because of their need for full sunlight and their inability to compete with hardwood species. The same was true of red cedar. In southern New England, abandoned croplands were more often than not invaded by gray birch; abandoned pastures, on the other hand, were taken over by red cedar and white pine.[31]

Livestock not only helped shift the species composition of New England forests but made a major contribution to their long-term deterioration as well. If colonial lumberers made sure that woods were stripped of their largest and oldest trees, grazing animals made sure that those trees were rarely replaced. Benjamin Lincoln wrote with some emotion when he argued:

> We suffer exceedingly at this day by the ill judged policy of permitting the cattle to run at large in the woods, especially in the full settled towns. Those tracts reserved for building, timber, fence-stuff, and fuel, are constantly thinning, and many of them are ruined as wood land, there are so large a proportion of cattle turned out, compared with the plants which come up in the spring, and the shoots which appear around the stumps of trees fallen the year before.

To Lincoln, allowing animals to graze in the woods was to let trees be "wantonly destroyed," and he sought to show that doing so was actually "more expensive and injurious to the common interest, than if lands were ploughed, and grain sowed, on which they might feed."[32]

Lincoln's concern was well-founded. Wherever the English animals went, their feet trampled and tore the ground. Because large numbers of them were concentrated on relatively small tracts of land, their weight had the effect of compacting soil particles so as to harden the soil and reduce the amount of oxygen it contained. This in turn curtailed the root growth of higher plants, lowered their ability to absorb nutrients and water, and encouraged the formation of toxic chemical compounds. Soil compaction, in other words, created conditions that were less hospitable to plant life and eventually lowered the soil's carrying capacity for water. (One of the things that distinguished European clover and timothy grass from other plants was precisely their ability to live on severely compacted soils containing little oxygen.) Ironically, then, an additional effect of woodland grazing was to kill many of the plants on which livestock depended for food, so that animals ran out of browse before their grazing season was over. Their survival in these circumstances depended on the colonists' efforts to open new pastures, create additional hay meadows, or cultivate more grain crops. Pasture deterioration was thus an incentive for still more intensive colonial deforestation.[33]

But the greatest effect of domesticated animals on New England soils came in the one area from which they were systematically excluded during most seasons of the year: croplands. Precolonial Indian women had had only their hoes and their own hands to turn the soil; the colonists, on the other hand, could use their oxen and horses to pull plows, which stirred the soil much more deeply. Plowing destroyed all native plant species to create an entirely new habitat populated mainly by domesticated species, and so in some sense represented the most complete ecological transformation of a New England landscape. Animals made it possible for a single colonial family to farm much larger areas than their Indian predecessors had done. Moreover, colonial farmers, because of their fixed notions of property ownership, continued to plow the same fields years after Indians would have abandoned them. The intimate connection between grazing animals, plows, and fixed property lay at the heart of European agriculture, with far-reaching ecological consequences.[34]

Whatever the causes that reduced the ground cover of New England soils, the long-term effect was to put those soils in jeopardy. The removal of the forest, the increase in destructive floods, the soil compaction and close-cropping wrought by grazing animals, plowing—all served to increase erosion. The naturalist John Bartram wrote to Jared Eliot in the mid-eighteenth century and spoke of a time

> above 20 years past when the woods was not pastured and full of high weeds and the ground light[,] then the rain sunk much more into the earth and did not wash and tear up the surface (as now). The rivers and brooks in floods would be black with mud but now the rain runs most of it off on the surface[,] is colected into the hollows which it wears to the sand and clay which it bears away with the swift current down to brooks and rivers whose banks it overflows.

Though he wrote of the mid-Atlantic colonies rather than New England, Bartram described processes which were unquestionably going on in both regions. Within a year or two after a forest was cleared, its soil began to lose the nutrients that had originally sustained (and been sustained by) its ecological community. Particles of inorganic matter in its runoff water increased perhaps five- or sixfold, and dissolved minerals also washed away more quickly. In pastures and meadows, both effects were aggravated by the presence of grazing animals; in planting fields, deeply stirred soils came into greater contact with both air and water, thus decomposing organic material and losing dissolved nutrients more rapidly. The result was to reduce still further the ability of soils to sustain plant life.[35]

As more and more particulate matter entered stream courses, sedimentation rates in the various parts of New England watersheds began to change. Deposition of sediment in ponds and lakes probably jumped by a third or more shortly after European settlement began in an area, increasing gradually thereafter until such bodies of water were filling up as much as five times more quickly than they had in the precolonial period. The greater violence and shorter duration of spring runoffs meant that soil which had once been deposited along the banks of streams and rivers was no longer left there in such quantities, except in places where the current slowed. This meant that the interval lands once sought so eagerly for their rich soils by the first settlers were no longer so replenished by their annual inundations. Instead, the more violent floods, moving across lands disturbed by plow and animal alike, actually carried more soil away, and replaced it with poorer soil from upstream areas that were losing the upper layers of their topsoil. The fertility of at least some interval lands had begun to decline by the mid-eighteenth century, an effect which the agricultural writer Samuel Deane attributed to the more rapid subsiding of floods, "owing to the more cultivated state of the country."[36]

The city of New Haven furnishes a good example both of the drying and the erosion which accompanied English agricultural practices. The earliest map of the town, dated 1641, shows a substantial stream running across the east corner of the original plat, and a smaller one which flowed from the south corner of the town green. By 1724, both streams had vanished from the original plat, although they continued to run to the southeast of it. By 1802, the small stream had disappeared entirely, and the flow of the larger one had so declined that it was labeled a "canal." Neither exists at all today. The silt which had helped fill these streams also made its contribution to the shallow harbor upon which the city depended for shipping. Between 1765 and 1821, New Haven was forced to extend its main wharf more than 3900 feet into its bay to stay ahead of the mud which would otherwise have prevented ships from landing. And yet, as Dwight reported, there was "less water a few rods from its foot" in 1821 than there had been at the end of the much shorter wharf in 1765. Similar harbor changes were recorded in such places as Boston, Barnstable Bay, and Nauset, and although the danger of *post hoc ergo propter hoc* is here particularly acute, it seems reasonable that at least some of these alterations were encouraged by human agency. As in the case of colonial mills, shifting rates of stream flow, sedimentation, and erosion could all have unexpected consequences for local economies, and could change farms, pastures, and harbors in subtle but significant ways.[37]

Grazing animals and plowing brought especially severe erosion problems to beaches and areas with sandy soils. Dwight told of visiting a place near Saybrook, Connecticut, the soil of which had begun to blow off some forty years earlier, presumably because cattle had cropped the ground cover too closely. In the course of half a century, the soil of twenty acres had been exposed by wind erosion to a depth of one to three feet. The sand plains near North Haven, Connecticut, had been affected in a similar way. But perhaps the most interesting case of cattle-induced wind erosion occurred at Truro on Cape Cod. Cattle had so seriously damaged the beach grass there by the 1730s that sand was blowing into dunes and encroaching upon the town's meadows. The Massachusetts General Court therefore passed an act in 1739 forbidding all grazing of cattle in the affected areas. The law was repeated many times, and, each April, inhabitants were required as one of their services to the town to plant beach grass on all land whose sandy soil had become exposed. In some areas, such measures proved remarkably effective; in others, it was already too late. An anonymous traveler in 1794 told how "large tracts of land" in Truro "have now become unfit for cultivation." According to this visitor, however, Truro had "no such appearances of desolation, as are exhibited on the plains of Eastham, [also on Cape Cod,] where an extensive, and what was once a fertile spot, has become prey to the winds, and lies buried under a barren heap of sand." Offshore islands—favorite places for cutting wood and grazing cattle—were similarly at risk from this kind of erosion, and changed their shape or disappeared altogether in consequence.[38]

Deforestation, grazing, plowing, erosion, and watershed

changes all contributed to a problem that became endemic to colonial agriculture in New England: soil exhaustion. Lands cleared for crops frequently had to be turned back to pasture or woods less than a decade after their first planting. In this, colonial farmers were not radically different from their Indian predecessors: Indians too had moved their fields from place to place. But colonists tried to incorporate Indian practices into a much different system of agriculture and property boundaries, a system that led to more intensive land use and greater ecological change. They hastened soil exhaustion by practicing monoculture—raising corn without the accompanying legumes which had helped fertilize Indian fields—and also by letting their livestock eat cornstalks and other unharvested material which could have been plowed back into the soil. By removing such organic materials from the field, colonial farmers lost nutrients which the Indians had retained. Moreover, once colonists abandoned a field, they never let it lie wholly fallow but used it for pasture instead.

Part of the reason planting fields lost their fertility so quickly was the colonists' use of maize as a staple crop. Except for rice —a much more labor-intensive plant—no other grain produces so great a yield, and none puts greater demands on soil. The author of *American Husbandry* wrote in 1775: "Maize is a very exhausting crop; scarce any thing exhausts the land more." Because its yield was so great, and because Old World grains were more difficult to raise, maize quickly came to be used by colonial farmers with little rotation of crops. European plow agriculture was bested suited to grains like wheat and rye, which could be sowed by scattering to create a dense groundcover; maize, which Indians had developed for hoe agriculture, grew best when planted individually and so covered relatively little ground. It thus not only consumed large quantities of nutrients in its own right but also did little to stop unused nutrients in plowed soil from washing away. The resulting soil exhaustion could have been predicted, and was less the fault of the soil than of the husbandry. By 1637, one disillusioned planter wrote that the soil "after five or six years . . . grows barren beyond belief; and . . . puts on the face of winter in the time of summer." Without fertilizer, such soil became useless for crops.[39]

Unfortunately, New England farmers, in allowing their livestock to wander over much of the landscape, lost one of European agriculture's principal sources of fertilizer: animal manure. Colonists did use what little dung collected in their farmyards, and some communities even created a market whereby the town's flock of sheep spent their nights on the fields of whoever bid the most for them. By this means, as Sarah Knight explained, "they will sufficiently Dung a Large quantity of Land before morning." But such schemes were relatively rare, and most manure was simply lost. No concentrated manuring could be done to increase crop yields because cattle were rarely housed at night in a place where their dung could be gathered. By failing to collect manure, colonists contributed to the exhaustion not only of croplands but of meadows and pastures as well. New England farmers rarely concentrated their hay crop into a small area, with the result that they mowed the greater portion of their land in the autumn, however thin the grass. This left the soil almost devoid of ground cover during the winter, promoted erosion, and so yielded smaller hay crops each year. Smaller crops meant a reduced store of hay each winter. As supplies ran low, farmers were forced to turn their livestock out to graze as early as possible in the spring, and the animals obliged by eating new shoots of grass almost before they were out of the ground. Their hooves pockmarked and compacted the wet soil, destroyed roots, encouraged soil loss, and made future mowing still more difficult. The vicious cycle could be broken only by keeping mowing land well separated from grazing land and making sure that it was properly manured. Unfortunately, "the generality of farmers" did not do so, and continued to rely on hay rather than cabbages, turnips, and other root crops as their principal source of fodder.[40]

Because no manure could be gathered from livestock which were not housed in barns at night, the English turned to fish as an alternate source of fertilizer. Whereas Indians had fished the spring spawning runs primarily for their own food, the colonists did so in order to apply tens of thousands of alewives, menhaden, and other fish to their cornfields. As early as 1634, a market in fish fertilizer was already in existence in Massachusetts, and was still going strong in some areas at the end of the eighteenth century, when a thousand fish could be had for a dollar. The fish did a remarkably good job of prolonging soil fertility, but had certain definite drawbacks. They attracted wild animals, who tried to dig them out of the mounds in which maize was planted, and in the long run the fish tended to spoil fields with their oiliness. For travelers, their most distressing characteristic was the "almost intolerable fetor" with which they filled the air; farmers seemed to grow used to it, but Dwight declared it "extremely disgusting to a traveler."[41]

Fish fertilizer carried problems of supply as well. It was available primarily to farmers living in towns along the coast, and to those living on the rivers and streams which served as spawning routes. Unfortunately, these latter became less common as time passed. Especially toward the end of the eighteenth century, the building of dams for mills and canals prevented not only alewives but also larger fish from returning upstream. Whole rivers were depopulated as a result: salmon were gone from the Piscataqua by 1750, and by the early nineteenth century had departed from the upper reaches of the Farmington, the Connecticut, and the Salmon Fall Rivers as well.[42]

Farmers without fish had to resort to other methods. Those lucky enough to live on the annually flooded interval lands at least for a while seemed blessed with perpetual fertility. Others tried to solve the problem of what one called "our old land which we have worn out" by applying ashes, a method which destroyed forests for the benefit of fields. Jared Eliot said, "*Ashes* is allowed on all hands to be some of the best Dressing or Manure for land; it inricheth much and lasts long." For colonial farmers in the eighteenth century, it was almost proverbial that "if we could get a sufficiency of Ashes, we could do well enough." Unfortunately, as Eliot pointed out, "the misery is we can get but little," and it took "a great deal of Wood to make a little Ashes." The only other solution was to seed the land with nitrogen-fixing clover and mow it for several years before trying to plant it again. Crop rotation was not well understood by most colonial farmers, however; turning the land to fallow and grazing it was probably the fate of most exhausted fields. Depending on the intensity with which they were grazed or mowed, it might take many years before they recovered their fertility, and some never did. Dwight told of a forsaken and windblown field near Truro which had finally been abandoned as barren. "Yet," he wrote, "these lands are said in ancient times to have produced fifty bushels of maize to the acre, and from fifteen to twenty bushels of wheat." As in the process of clearing, colonial farmers treated their land as a resource to be mined until it was exhausted, rather than one to be conserved for less intense but more perennial use.[43]

Soil exhaustion was not the only result of reproducing European farming practices in New England. Indian fields, which in addition to being both small and diverse were subjected to regular burning, had originally been troubled by relatively few diseases and pests. The arrival of the colonists changed this. By using their animals and plows to create more extensive areas of cropland than the Indians had done, colonists unintentionally created habitats which many organisms found quite attractive. Accordingly, when colonists brought their new crops and livestock to America, not all the weeds which joined them on the journey were plants. A number of the unintended migrants were animals. Unlike the colonists' plant weeds, however, they thrived not on disturbed soils that were stripped of vegetation but on soils whose native plants had been replaced by concentrated food plants.

One of the more dramatic of these "animal weeds" was the Hessian fly, which made its first American appearance on Long Island during the Revolutionary War. Supposedly brought to the New World by the Hessian mercenaries from whom it derived its name, it rapidly spread to New Jersey and Connecticut and proceeded to lay waste to entire wheatfields. By the end of the century, it had brought about the virtual end of wheat raising in Connecticut, completely extirpating the white bald wheat which had made up the bulk of colonial production. Timothy Dwight told of a merchant in West Greenwich who claimed that "the inhabitants of that parish before the arrival of the fly used to export annually ten thousand bushels of wheat, but were then obliged to import three thousand." Farmers eventually learned to use different varieties of wheat and to sow them just before the first frost, when adult flies would be killed by the cold, but New England wheat production was never the same again.[44]

Other animal imports were less destructive than the Hessian fly, however annoying they might have seemed to the colonists. Among these were the ordinary black fly and that perennial occupant of Old World houses, the cockroach. A more benign insect migrant was the honeybee, which was carried to the colonies as a domesticate and rapidly became established in the wild. There were even mammalian weeds. The gray rat was a regular shipboard traveler which spread inland from port cities as English agricultural settlement proceeded. The house mouse did likewise.[45]

But not all animal weeds were imports. Monocultural crop concentration encouraged its own pests. Native species which ordinarily fed on grain or similar plants suddenly found their food supplies copiously augmented by colonial farmers—in fields that were larger and closer together—with the result that their populations sometimes increased dramatically. One of the subtler effects of Indian burning had been to hold such populations in check; without it, agricultural pests became a more and more serious problem. An invasion of caterpillars assaulted New England grain crops in 1646, and the animals reappeared with some regularity thereafter. Other insects that increased their populations by inhabiting colonial gardens included grasshoppers, garden fleas, maggots, and various species of "worms" and "flies." Colonial orchards were particularly susceptible to attack by such insects, and, in the case of the cankerworm, which first appeared in 1666, had to be constantly defended against them. The concentration of animal food supplies similarly increased populations of squirrels and crows, which became so common in the vicinity of colonial fields that, as with wolves, bounties were offered for their skins. As Peter Kalm explained in relation to squirrels, "The infinitely greater cultivation of corn, which is their favorite food, is the cause of their multiplication." The native field mice also rose to prodigious numbers. All of these creatures served to complicate the very agriculture which had made possible their proliferation.[46]

But the most serious threat to English crops, especially wheat and rye, was not an animal but a fungus: the "blast," or black stem rust, an Old World disease which first appeared in New England in the early 1660s. Because the New England climate was already marginal for wheat raising, the blast's immediate effects were devastating. John Hull recorded in his diary that "the wheat throughout our jurisdiction this year mostly blasted: in sundry towns, scarce any left." The disease destroyed the leaves of wheat plants before they had a chance to produce seed, and it became a perennial problem for the rest of the colonial period. It resulted in the virtual elimination of wheat raising in a number of settlements, especially in older ones which had been

farmed for a number of years. "The great discouragement" of
New England farmers, wrote Thomas Hutchinson, "has been
the blast." Colonists soon discovered that the blight was most
common in areas where barberry bushes—another imported
European weed, one that particularly favored the edges of pastures—were growing, and began to lobby for the destruction of
these plants. In response, Connecticut in 1726 passed a law calling
for the destruction of all barberries, and Massachusetts and
Rhode Island passed similar laws later in the century. They
proved decidedly ineffective—Dwight viewed it as "altogether
improbable . . . that they [the barberries] will ever be extirpated"
—but the ecological insight which had produced the laws was
indubitably correct. Barberries were indeed the host which supported one phase of the rust's life cycle, and so produced thousands of spores that destroyed any wheat plants which lay down-
wind of them. A European weed, in other words, had brought
with it a European disease that made it exceedingly difficult for
European farmers, keeping European animals, to raise a key
European crop. The blasting of wheat was thus a kind of metaphor for the extent to which Old World ecological relationships
had been reproduced in New England.[47]

Toward the end of the eighteenth century, other colonial activities began to have significant effects on New England ecosystems. The draining of swamps and salt marshes became more
frequent as greater amounts of capital were invested in agriculture. In the same way, it became more common for farmers to
irrigate meadowlands to increase hay yields. Dams for irrigation,
millponds, and canals brought with them significant changes in
fish populations, and to some extent helped reverse the drying of
the land which had begun with the destruction of beaver dams.
Mosquito-borne illnesses sometimes became more frequent in
the newly flooded areas behind such dams, as did flooding during
the spring runoff. The building of iron furnaces in Rhode Island
and in the mountains of western Massachusetts and Connecticut
raised the fuel consumption of those regions considerably. Each
ton of finished iron required the burning of 250 bushels of charcoal, which in turn could only be obtained by burning a pile of
wood 4 by 4 by 48 feet. Since a single furnace might produce well
over five hundred tons of iron in a year, the growth of the iron
industry significantly increased the rate at which local forests

were cut for fuel. Tanneries also became more common, encouraging the selective cutting of oak and hemlock for their tanbark.[48]

All these industrial activities, however, were really only beginning by the end of the colonial period. The Industrial Revolution and the national expansion of American agriculture would again transform New England ecology during the nineteenth century. Industry would concentrate populations in urban centers, and these would become the principal markets for a local agriculture that—save for orchards, dairies, and market gardens—was increasingly in decline. The opening of the Erie Canal, and later the railroads, would connect New England cities to the grain-raising and meat-producing regions of the Middle West, and these would deliver the final blow to farms that were already becoming marginal for both economic and ecological reasons. Southern New England was as much as three-fourths deforested at the middle of the nineteenth century; since that time, forests and cities have continuously encroached on cleared farmlands, so that the bulk of them no longer stand as fields. New England's stone walls now wander through woods rather than pastures. Indeed, so complete have been the urban-industrial transformations of the region that it is now difficult to imagine the extent to which the earlier colonial landscape itself represented a transformation of its Indian predecessor. But such, in fact, it was. The colonial interaction of forests, furbearers, hunters, axes, grazing animals, plows, crops, weeds—and the rival ways of owning and selling these things—all contributed to a redrawn map of New England. It was a map that, over the course of European settlement, more and more traced, not the earlier world of movement between hunt and harvest, but the new world of cropland and pasture, of agricultural cycles entrapped within the fixed boundaries of individual possession. In the hands of the colonists, New England had become a world of fields and fences.

PART III

Harvests of Change

8

THAT WILDERNESS
SHOULD TURN A MART

New England in 1800 was far different from the land the earliest
European visitors had described. By 1800, the Indians who had
been its first human inhabitants were reduced to a small fraction
of their former numbers, and had been forced onto less and less
desirable agricultural lands. Their ability to move about the landscape in search of ecological abundance had become severely
constrained, so that their earlier ways of interacting with the
environment were no longer feasible and their earlier sources of
food were less easy to find. Disease and malnutrition had become
facts of life for them.

Large areas particularly of southern New England were now
devoid of animals which had once been common: beaver, deer,
bear, turkey, wolf, and others had vanished. In their place were
hordes of European grazing animals which constituted a heavier
burden on New England plants and soils. Their presence had
brought hundreds of miles of fences. With fences had come the
weeds: dandelion and rat alike joined alien grasses as they made
their way across the landscape. New England's forests still ex-
ceeded its cleared land in 1800, but, especially near settled areas,

the remaining forest had been significantly altered by grazing, burning, and cutting. The greatest of the oaks and white pines were gone, and cedar had become scarce. Hickory had been reduced because of its attractiveness as a fuel. Clear-cutting had shifted forest composition in favor of those trees that were capable of sprouting from stumps, with the result that the forests of 1800 were physically smaller than they had been at the time of European settlement. The cutting of upland species such as beech and maple, which were accustomed to moist sites, produced drying that encouraged species such as the oaks, which preferred drier soils.

Deforestation had in general affected the region by making local temperatures more erratic, soils drier, and drainage patterns less constant. A number of smaller streams and springs no longer flowed year-round, and some larger rivers were dammed and no longer accessible to the fish which had once spawned in them. Water and wind erosion were taking place with varying severity, and flooding had become more common. Soil exhaustion was occurring in many areas as a result of poor husbandry, and the first of many European pests and crop diseases had already begun to appear. These changes had taken place primarily in the settled areas, and it was still possible to find extensive regions in the north where they did not apply. Nevertheless, they heralded the future.

Why had these things happened?

To compare New England ecosystems in 1600 with those in 1800 as if examining two snapshots—New England before the Europeans and New England after—is to imply that the European invasion was the chief agent of environmental change. In a crude sense, there can be little doubt that this was true. Most of the transformations described above would not have occurred had the Atlantic never been crossed, and so our analysis of ecological change must inevitably focus on differences between the human communities that existed on opposite sides of the ocean: differences in political organization, in systems of production, and in human relationships with the natural world. The shift from Indian to English dominance in New England saw the replacement of an earlier village system of shifting agriculture and hunter-gatherer activities by an agriculture which raised crops and domesticated animals in household production units that were contained within fixed property boundaries and linked with commercial markets. Ultimately, English property systems encouraged colonists to regard the products of the land—not to mention the land itself—as commodities, and so led them to orient a significant margin of their production toward commercial sale in the marketplace. The rural economy of New England thus acquired a new tendency toward expansion. The dynamics which led colonists to accumulate wealth and capital were the most dramatic point of contrast between the New England economy of 1600 and that of 1800. The economic transformation paralleled the ecological one, and so it is easy to assert that the one caused the other: New England ecology was transformed as the region became integrated into the emerging capitalist economy of the North Atlantic. Capitalism and environmental degradation went hand in hand.

And yet the problem is not quite so simple. One serious danger of a two-point analysis which contrasts New England before and after the Europeans is that it obscures the actual processes of ecological and economic change. It makes that change seem too sudden and unicausal. The Europeans brought to the New World, not just new economic institutions, new markets, and new ways of bounding the landscape, but other things that are less easy to attribute to the direct agency of "capitalism." The devastating effects, for instance, of the disease organisms which wrought such havoc with Indian populations were primarily a function of the Indians' isolation from Old World disease environments, and would have been similar no matter what the economic organization of the European invaders. For the Indians, new diseases were one of the clearest consequences of European settlement, but once present, their effects had more to do with biology than economics. This is not to say, of course, that biology and economics were unrelated. That sailors and settlers came to America in sufficient numbers to bring diseases with them was a direct result of social and economic transformations in Europe. By the same token, the demographic collapse which diseases visited upon Indian populations was instrumental in disrupting the Indians' status systems so as to encourage their participation in the fur trade; diseases also had the effect of clearing the land of its earlier inhabitants and facilitating its conquest by European settlers. If Europeans were responsible for bringing dis-

eases to America, it is no less true that those diseases in turn
helped promote European expansion. They were as much a socio-
economic fact as an ecological one.

Similar claims about multiple causation can be made for many
of the ecological relationships which English settlers eventually
reproduced in New England. As we have seen, domesticated
animals exercised a profound influence on New England landscapes, and represented a dramatic contrast between the ways
Europeans and Indians went about obtaining their livelihoods.
But here again there are dangers in attributing the effects of
livestock solely to capitalist expansion: Old World pastoralism
antedates capitalism by four or five thousand years. Livestock—
whether raised for market or for home consumption—were
themselves a major reason for the dispersal of colonial settlements. Ecological pressures brought on by overgrazing and
inadequate forage reinforced economic incentives flowing more
directly from market demand: together, the two impelled colonial movement onto new lands. Similar arguments can be applied
to European grain production and forest clearing. Insofar as
these practices would have provoked ecological changes no matter what the economy in which they were embedded, their effects
cannot be attributed only to the expanding markets and trade
relationships of European capitalism—and yet it was precisely
those markets and relationships, themselves being transformed
in the transition to capitalism, which had brought Europeans to
America in the first place. Economic and ecological imperialisms
reinforced each other.

No one was more aware of this fact than the Indians. One of
the most perceptive analyses of ecological change in early New
England was delivered in a speech by the Narragansett sachem
Miantonomo in 1642, just a few years after English colonists
began to settle in the vicinity of his people's villages. "You
know," he said, speaking of a time just recently past,

> our fathers had plenty of deer and skins, our plains were
> full of deer, as also our woods, and of turkies, and our coves
> full of fish and fowl. But these English having gotten our
> land, they with scythes cut down the grass, and with axes
> fell the trees; their cows and horses eat the grass, and their
> hogs spoil our clam banks, and we shall all be starved.

For Miantonomo, as for other New England Indians, the struggle against the English, in its most basic form, was a result of the colonists "having gotten our land." The English accomplished this by a wide variety of means: their different conceptions of property, their willingness to use military force and legal deceit in acquiring land, their ideology of conquest and conversion, and so on. The Indian response to English land hunger was a shrewd mixture of economic self-interest and cultural adjustment, but was ultimately expressed as political resistance. Many villages which had initially welcomed the English presence as a means of acquiring trade goods and making powerful allies eventually chose to fight further colonial encroachments on their territories. They did so by forging new alliances with other Indian (and European) groups, responding with great creativity to the new diplomatic circumstances in which they found themselves. Miantonomo himself used his ecological analysis to argue for the necessity of a new Indian unity to match that of the English: "for so," he said, "are we all Indians as the English are, and say brother to one another; so must we be one as they are, otherwise we shall be all gone shortly." His argument for a new pan-Indian unity led him finally to propose that he and his allies ambush the colonists, "and kill men, women, and children, but no cows." The latter, he said, should be used for food "till our deer be increased again."[1]

Miantonomo's speech illustrates the sometimes contradictory ways in which Indians responded to the European threat: theirs was a flexibility whose range of choices was increasingly constrained by colonial dominance. As in the case of the European cows which Miantonomo thought might replace Indian deer, Indians were quite willing to adopt and modify European tools and textiles to their own purposes. They learned to use and repair European weapons, and oriented their hunting toward production for European markets. They began raising livestock, expanding the size of their corn crops, and practicing more sedentary ways of life. Their political communities became more extensive in the form of tribal alliances and confederacies in order to meet the need for pan-Indian unity and resistance which Miantonomo described. Indians, in other words, adjusted to what the Europeans brought to New England by modifying the ways they obtained their livelihoods, but at the same time retained their political and cultural identity. By ceasing to live as their ancestors had done, they did not cease to be Indians, but became Indians with very different relationships to the ecosystems in which they lived. Only in this limited—but ecologically crucial —sense can we say that an earlier Indian way of life had become impossible by 1800: although the subsistence practices of the New England Indians resembled those of European peasants more than they had before, Indians continued to define themselves as people apart, people resisting full incorporation into the world of their conquerors. The material conditions which had allowed them to practice their annual journey through the seasons no longer existed, but the Indians themselves remained, however much their communities and economies had changed with their environment.[2]

There is thus a second danger in analyzing New England ecological change simply by contrasting two landscapes, one before, the other after, the Europeans arrived. By making the arrival of the Europeans the center of our analysis, we run the risk of attributing all change to their agency, and none to the Indians. The implication is not only that the earlier world of "Indian" New England was somehow static but also that the Indians themselves were as passive and "natural" as the landscape. In fact, the Indians were anything but passive in their response to European encroachments. Faced with what they perceived as new opportunities, they took them as they saw fit; faced with threats to their political autonomy, they fought back as best they could. There is no inherent reason to believe that Indians could not have made far more dramatic adjustments than they did to their new ecological circumstances, if in no other way than by becoming fuller participants in the North Atlantic economy. That they generally did not do so must be attributed in part to their own choice and in part to the English refusal—whether enforced by violence or by law—to let them do so. If Indian communities were no longer autonomous political entities by 1800, it was because English colonists had made them so, denying them access to the land and resources which would have allowed them a more independent existence.

Instead, those Indians who remained in New England were confined to reservations, forced onto inferior farmlands, left without animals to hunt or fish, and so had to make their ecological adjustments in a far from ideal setting. Being overpowered is not a sign of passivity: however large the ecological forces that drove Indian communities toward change, it is crucial that we see those forces in their political and economic context. Ecology can help us analyze why Indians in 1800 had trouble sustaining themselves on the lands which remained to them, but it cannot explain why they had been compelled to live on those lands in the first place. Only politics can do that. Here the fate of Miantonomo can serve as a token of the political interactions between colonists and Indians which accompanied the ecological conflicts he described so well in his speech. Ransomed by his tribe from a rival sachem who had captured him, he requested that he be turned over for safekeeping to the English with whom he was then allied. The colonists responded by arranging for his assassination: he was murdered in cold blood.[3]

Putting the ecological transformations of colonial New England in their larger political and economic context carries us full circle to the expansion of European capitalism in the seventeenth and eighteenth centuries. Even after we have admitted the multicausal quality of the European institutions transferred to the New World, even after we have acknowledged the autonomous agency of the Indians in meeting the challenges those institutions posed, we are still confronted with a regional ecology which in 1800 bore fundamentally new relationships to other parts of the world. Those new relationships had as their source a new human perception of how the resources of the New England landscape might be made useful to those who could possess them. As the French anthropologist Maurice Godelier has remarked, a natural "resource" cannot exist without some intervening human agency which defines it: "there are thus," he writes, "no resources as such, but only possibilities of resources provided by nature in the context of a given society at a certain moment in its evolution." By drawing the boundaries within which their exchange and production occur, human communities label certain subsets of their surrounding ecosystems as resources, and so locate the meeting places between economics and ecology.[4]

All communities exercise choice in their labeling of resources, but they do so in radically different ways. Perhaps the central contrast between Indians and Europeans at the moment they first encountered each other in New England had to do with what they saw as resources and how they thought those resources should be utilized. Indians had a far greater knowledge of what could be eaten or otherwise made useful in the New England environment; their economy defined a correspondingly greater range of resources. But most of those resources were simply used or consumed by the household which acquired them, or, if exchanged, were traded for similar items. Very few resources were accumulated for the explicit purpose of indicating a person's status in the community: wampum, furs, certain minerals, and ornaments of the hunt generally served these purposes. Class authority was maintained more by kin networks and personal alliances than by stores of wealth, and the latter were in any event limited by the community's commitment to geographical mobility. There was thus little social incentive to accumulate large quantities of material goods. A wide range of resources furnished economic subsistence, while a narrow range of resources conferred economic status. The community's social definition of "need" was inherently limited, and made economic abundance a relatively easy attainment for its members. It was for this reason that Roger Williams could write of the Narragansetts: "Many of them naturally Princes, or else industrious persons, are rich; and the poore amongst them will say, they want nothing." Rich and poor alike were relatively easily satiated, and so made relatively slender demands on the ecosystems which furnished their economy its resources.[5]

The same could hardly be said of the European colonists. For them, perceptions of "resources" were filtered through the language of "commodities," goods which could be exchanged in markets where the very act of buying and selling conferred profits on their owners. Because European economies measured many more commodities in terms of money values—abstract equivalencies which could be accumulated, no matter what the resource involved, to become indicators of wealth and social status—they had few of the limitations which constrained the growth of their Indian counterparts. As a result, European markets, as the anthropologist Marshall Sahlins has suggested, at least in theory "erected a shrine to the Unattainable: *Infinite Needs.*" Those needs were determined not only by the local communities which became established in colonial New England but by all the distant places to which those communities sold their goods. The landscape of New England thus increasingly met not only the needs of its inhabitants for food and shelter but the demands of faraway markets for cattle, corn, fur, timber, and other goods whose "values" became expressions of the colonists' socially determined "needs." Ironically, though colonists perceived fewer *resources* in New England ecosystems than did the Indians, they perceived many more *commodities,* and so committed much wider portions of those ecosystems to the marketplace. "Nor could it be imagined," wrote the colonial historian Edward Johnson in 1653, "that this Wilderness should turn a mart for Merchants in so short a space, Holland, France, Spain, and Portugal coming hither for trade."[6]

The process whereby colonists (as well as Indians) linked New England ecosystems to market relationships was neither instantaneous nor continuous. Just because the earliest English explorers perceived the New England coast in terms of its commodities does not mean that their perceptions had immediate ecological consequences. Colonial economies underwent nearly as profound an evolution in New England as those of the Indians. Many English colonists, for example, initially supplemented their agriculture with subsistence activities—hunting and gathering—which looked distinctly Indian; colonists were eventually forced to limit these for the same ecological reasons that Indians did. Colonial systems for fixing property boundaries were not fully articulated until late in the seventeenth century. The degree to which land was committed to commercial production depended upon a host of changing factors: population growth, imperial regulations, ease of transportation to urban markets, and so on. Most early farmers directed only a small margin of their production to market sale. Perhaps their most important attachment to the market was not even related to immediate production —their expectation that the size of that margin would increase, and with it the value of their land. The inhabitants of some New England towns speculated in land almost from the beginning, but others awaited market developments of the late seventeenth and the eighteenth centuries before real estate was treated as so abstract a commodity. Markets, in other words, like commodities, were socially defined institutions which as a result of the transition to capitalism operated very differently at the end of the colonial period than they had at its beginning.

However true this may be, it must nevertheless be repeated that the abstract concept of the commodity informed colonial decision-making about the New England environment right from the start. The colonists brought with them concepts of value and scarcity which had been shaped by the social and ecological circumstances of northern Europe, and so perceived New England as a landscape of great natural wealth. Searching for commodities which would allow them to obtain European goods, they applied European definitions of scarcity—that is to say, European prices—to New England conditions of abundance. Operating in an economy where labor was scarce and difficult to hire, where accumulated capital was smaller than it had been in Europe, colonists turned to the factor of production which could compensate for the ones they lacked: they turned to the land and all it contained. Fish, fur, and lumber were assigned high values because of their scarcities in Europe, but were more or less free goods in New England. They had only to be taken and transported to market to yield a substantial return on invested labor; because of this, they were treated as wasting assets capable of rapid conversion to more liquid capital. Labor cost alone operated as a constraint on their exploitation, since colonists could consume natural wealth as a substitute for capital.

The result was an economy which used natural resources in a way which often appeared to European visitors as terribly wasteful. "In a word," wrote the Swedish traveler Peter Kalm of American farming practices, "the grain fields, the meadows, the forests, the cattle, etc. are treated with equal carelessness." A number of Americans agreed. In 1787, the physician Joseph Warren wrote a critique of American agriculture in which he argued:

> There is, perhaps, no country in the world, where the situations, nature, and circumstances of things, seem to point out husbandry as the most essential and proper business, more than our own; and yet, there is scarcely one where it is less attended to.

Warren attributed this apparent paradox to several factors: the Americans' tendency to farm overlarge tracts of land, their "rage for commerce," their investment of little capital in their farmlands, and their wasteful practices in feeding livestock. At the most basic level, however, what distinguished European and American farms was their production of nearly identical commodities with very different proportions of labor and land. As Warren noted, "Nothing will give a clearer idea of the different management, than the following facts: in England, rents are high and labour low; in America, it is just the reverse, rents are low and the rate of labour high." Here there was no paradox: American relations of production were premised upon ecological abundance, and so attached a higher value to labor than had been the case in Europe. Returns to labor were so high in America precisely *because* returns to land were so low.[7]

Land in New England became for the colonists a form of capital, a thing consumed for the express purpose of creating augmented wealth. It was the land-capital equation that created the two central ecological contradictions of the colonial economy. One of these was the inherent conflict between the land uses of the colonists and those of the Indians. The ecological relationships which European markets created in New England were inherently antithetical to earlier Indian economies, and so those economies were transformed—as much through the agency of the Indians as the Europeans—in ways that need not be repeated here. By 1800, Indians could no longer live the same seasons of want and plenty that their ancestors had, for the simple reason that crucial aspects of those seasons had changed beyond recognition.

But there was a second ecological contradiction in the colonial economy as well. Quite simply, the colonists' economic relations of production were ecologically self-destructive. They assumed the limitless availability of more land to exploit, and in the long run that was impossible. Peter Kalm described the process whereby colonial farmers used new land until it was exhausted, then turned it to pasture and cut down another tract of forest. "This kind of agriculture will do for a time," he wrote, "but it will afterwards have bad consequences, as every one may clearly see." Not only colonial agriculture, but lumbering and the fur trade as well, were able to ignore the problem of continuous yield because of the temporary gift of nature which fueled their continuous expansion. When that gift was finally exhausted, ecosystems and economies alike were forced into new relationships: expansion could not continue indefinitely.[8]

The implications of this second ecological contradiction stretched well beyond the colonial period. Although we often tend to associate ecological changes primarily with the cities and factories of the nineteenth and twentieth centuries, it should by now be clear that changes with similar roots took place just as profoundly in the farms and countrysides of the colonial period. The transition to capitalism alienated the products of the land as much as the products of human labor, and so transformed natural communities as profoundly as it did human ones. By integrating New England ecosystems into an ultimately global capitalist economy, colonists and Indians together began a dynamic and unstable process of ecological change which had in no way ended by 1800. We live with their legacy today. When the geographer Carl Sauer wrote in the twentieth century that Americans had "not yet learned the difference between yield and loot," he was describing one of the most longstanding tendencies of their way of life. Ecological abundance and economic prodigality went hand in hand: the people of plenty were a people of waste.[9]

NOTES

1. The View from Walden

1. Henry David Thoreau, *The Journal of Henry D. Thoreau,* Bradford Torrey and Francis H. Allen, eds., 2 vols. (original edition, 1906, New York, 1962), VII, pp. 132-7 (January 24, 1855).
2. *Ibid.,* VIII, pp. 220-1 (March 23, 1856).
3. Edward Johnson, *Johnson's Wonder-Working Providence,* J. Franklin Jameson, ed. (New York, 1910), p. 210; Benjamin Rush, *Essays, Literary, Moral and Philosophical,* 2nd ed. (Philadelphia, 1806), p. 221.
4. Timothy Dwight, *Travels in New England and New York* (1821), Barbara Miller Solomon, ed. (Cambridge, MA, 1969), IV, p. 186.
5. H. I. Winer, *History of the Great Mountain Forest, Litchfield County, Connecticut,* Ph.D. Thesis, Yale University, 1955, pp. 98-9; Thomas G. Siccama, "Presettlement and Present Forest Vegetation in Northern Vermont with Special Reference to Chittenden County," *American Midland Naturalist,* 85 (1971), pp. 153-72.
6. For a review of the literature using these techniques, see the bibliographical essay.
7. Marquis de Chastellux, *Travels in North America in the Years 1780, 1781 and 1782* (1786), Howard C. Rice, Jr., ed. (Chapel Hill, 1963), I, p. 78.
8. Winer, *Great Mountain Forest,* p. 78; Thoreau, *Journal,* VII, p. 133 (January 24, 1855); Chastellux, *Travels,* I, p. 78; Peter Kalm, *Travels in North America* (1753-61, 1770), Adolph B. Benson, ed., 2 vols. (New York, 1964), p. 50; J. Gordon Ogden III, "Forest History of Martha's Vineyard, Massachusetts: I. Modern and Pre-Colonial Forests," *American Midland Naturalist,* 66 (1961), p. 426; Stanley W. Bromley, "The Original Forest Types of Southern New England," *Ecological Monographs,* 5 (1935), pp. 72-6; Austin F. Hawes, "New England Forests in Retrospect," *Journal of Forestry,* 21 (March 1923), p. 209. On problems of animal nomenclature, see Frederick W. Warner, "The Foods of the Connecticut Indians," *Bulletin of the Archaeological Society of Connecticut,* 37 (1972), pp. 27-9.
9. Literally, "after this, therefore because of this." For a discussion, see David Hackett Fischer, *Historians' Fallacies* (New York, 1970), pp. 166-7.

10. Donald Worster, *Nature's Economy* (San Francisco, 1977), Part IV; Ronald C. Tobey, *Saving the Prairies* (Berkeley, 1981).
11. For examples, see Hugh M. Raup and Reynold E. Carlson, "The History of Land Use in the Harvard Forest," *Harvard Forest Bulletin*, 20 (1941), p. 59; George E. Nichols, "The Vegetation of Connecticut: II. Virgin Forests," *Torreya*, 13 (1913), pp. 199-215; for a critique, see Margaret B. Davis, "Phytogeography and Palynology of Northeastern United States," in H. E. Wright and David G. Frey, eds., *The Quaternary of the United States* (Princeton, 1965), pp. 382-5, 397.
12. The crucial essay in promoting the ecosystem concept was A. G. Tansley, "The Use and Abuse of Vegetational Concepts and Terms," *Ecology*, 16 (1935), pp. 284-307, which is well discussed in Tobey, *Saving the Prairies*, Chapter 6; see also H. A. Gleason, "The Individualistic Concept of Plant Association," *Bulletin of the Torrey Botanical Club*, 53 (1926-27), pp. 7-26.
13. Francis Jennings, *The Invasion of America* (Chapel Hill, 1975), p. 15.
14. For sources on ecological anthropology, which has engaged in some of the most interesting discussions of this nature/culture problem, see the bibliographical essay. The two analyses referred to in the text are Roy A. Rappaport, *Pigs for the Ancestors* (New Haven, 1968); and Marvin Harris, "The Cultural Ecology of India's Sacred Cattle," *Current Anthropology*, 7 (1966), pp. 51-9. For a historian who has tried to apply similar notions to an American Indian case study, see Calvin Martin, *Keepers of the Game* (Berkeley, 1978).
15. A critique of functionalism comparable to the one I offer here can be found in Jonathan Friedman, "Marxism, Structuralism and Vulgar Materialism," *Man*, N.S., 9 (1974), pp. 444-69. Rappaport's response to such criticisms is "Ecology, Adaptation and the Ills of Functionalism," *Michigan Discussions in Anthropology*, 2 (Winter 1977), pp. 138-90; Friedman's counter-response is quite suggestive for historians: "Hegelian Ecology: Between Rousseau and the World Spirit," in P. C. Burnham and R. F. Ellen, eds., *Social and Ecological Systems* (New York, 1979), pp 253-70. See also Maurice Godelier, "Anthropology and Economics," in Godelier, *Perspectives in Marxist Anthropology* (Cambridge, England, 1977), pp. 15-62.
16. Alfred W. Crosby, Jr., *The Columbian Exchange* (Westport, CT), 1972; W. H. McNeill, *Plagues and Peoples* (New York, 1976).
17. See Catherine S. Fowler, "Ethnoecology," in Donald L. Hardesty, *Ecological Anthropology* (New York, 1977), pp. 215-43, for an introduction to such methods.
18. Thoreau, *Journal*, VIII, p. 221 (March 23, 1856).

2. Landscape and Patchwork

1. My initial sentence notwithstanding, the geographer Carl Sauer makes the important point that, for all of the differences, there were

many similarities between New England and England: "It would be impossible, indeed, to cross an ocean anywhere else and find as little unfamiliar in nature on the opposite side." ("The Settlement of the Humid East," in *USDA Yearbook, Climate and Man* [Washington, D.C., 1941], p. 159.) Sauer refers primarily to the individual species that were present in New England, whereas my argument speaks most directly to the populations of those species and the ecological relationships they bore to one another.

2. James Rosier, "A True Relation of the Voyage of Captaine George Waymouth, 1605," in Henry S. Burrage, ed., *Early English and French Voyages* (New York, 1906), p. 366; Richard Hakluyt, *Discourse Concerning Western Planting* (1584), in E. G. R. Taylor, ed., *Original Writings and Correspondences of the Two Richard Hakluyts* (Hakluyt Society, 1935), II, pp. 211-327. For problems of how to interpret words like "commodities" and "profits," see the *Oxford English Dictionary*, and the fuller discussion of these issues in Chapters 4 and 5.

3. John U. Nef, "An Early Energy Crisis and Its Consequences," *Scientific American*, 237:5 (November 1977), pp. 140-51; Nef, *The Rise of the British Coal Industry* (London, 1932); Carl Bridenbaugh, *Vexed and Troubled Englishmen* (New York, 1968), pp. 64, 98-101, 149; Charles F. Carroll, *The Timber Economy of Puritan New England* (Providence, 1973), pp. 3-21.

4. Martin Pring, "A Voyage Set Out from the Citie of Bristoll, 1603," in Burrage, *Early Voyages*, p. 349.

5. Douglas R. McManis, *European Impressions of the New England Coast, 1497-1620*, University of Chicago Department of Geography Research Paper No. 139 (Chicago, 1972), pp. 116-33, is good on this theme.

6. Francis Higginson, *New-Englands Plantation* (1630), *Massachusetts Historical Society Proceedings*, 62 (1929), p. 311; John Brereton, "Briefe and True Relation of the Discoverie of the North Part of Virginia, 1602," in Burrage, *Early Voyages*, p. 331; William Wood, *New England's Prospect* (1634), Alden T. Vaughan, ed. (Amherst, 1977), p. 56; Thomas Morton, *New English Canaan* (1632), Charles F. Adams, ed., *Pubs. of the Prince Society*, XIV (Boston, 1883), p. 222; John Josselyn, *An Account of Two Voyages to New England* (1675), in *Massachusetts Historical Society Collections*, 3rd ser., 3 (1833), p. 273.

7. Wood, *Prospect*, pp. 50-2; Morton, *Canaan*, p. 193; Higginson, *Plantation*, pp. 313-14; Josselyn, *Two Voyages*, p. 277; A. W. Schorger, *The Wild Turkey: Its History and Domestication* (Norman, OK, 1966), pp. 3-18. Wood's statement that some had killed fifty birds "at a shot" does not mean with a single gunshot but in a single hunting location. He and other seventeenth-century writers used "shot" in analogy with fishing, where the word means not only the single cast of a net but also the place from which nets are cast. See the *Oxford English Dictionary*.

8. Josselyn, *Two Voyages*, p. 278; Wood, *Prospect*, p. 50; William Hammond to Sir Simonds D'Ewes, September 26, 1633, in Everett Emer-

son, ed., *Letters from New England* (Amherst, 1976), p. 111; Thomas Dudley to Lady Bridget, Countess of Lincoln, March 12 and 28, 1631/31, Emerson, *Letters,* p. 81; A. W. Schorger, *The Passenger Pigeon: Its Natural History and Extinction* (Madison, 1955), pp. 3-13.

9. Higginson, *Plantation,* p. 310; Morton, *Canaan,* p. 199; Hammond to D'Ewes, in Emerson, *Letters,* p. 111.

10. John Josselyn, *New-Englands Rarities Discovered* (1672), in *Transactions and Collections of the American Antiquarian Society,* 4 (1860), pp. 147-8; Wood, *Prospect,* pp. 32, 51; Morton, *Canaan,* p. 214; Higginson, pp. 312-13; Hammond to D'Ewes, in Emerson, *Letters,* p. 110; Philip J. Greven, Jr., *Four Generations* (Ithaca, 1970), pp. 24-26; James A. Henretta, *The Evolution of American Society, 1700-1815* (Lexington, MA, 1973), pp. 9-15; but on epidemics, see also John Duffy, *Epidemics in Colonial America* (Baton Rouge, 1953), and Thomas Dudley, in Emerson, *Letters,* pp. 72, 76.

11. Emmanuel Altham to Sir Edward Altham, March 1623/24, in Sidney V. James, Jr., ed., *Three Visitors to Early Plymouth* (Plimoth Plantation, 1963), p. 36; Wood, *Prospect,* p. 38; Higginson, *Plantation,* p. 314.

12. Lawrence C. Wroth, ed., *The Voyages of Giovanni de Verrazzano, 1524-1528* (New Haven, 1970), p. 139; Wood, *Prospect,* pp. 38, 59; Higginson, *Plantation,* p. 308; John Winthrop, *Winthrop's Journal,* James Kendall Hosmer, ed. (New York, 1908), I, p. 258.

13. Wroth, *Verrazzano,* p. 140; McManis, *European Impressions,* pp. 49-67, 90-115; Josselyn, *Two Voyages,* p. 256; Morton, *Canaan,* p. 185; Howard W. Lull, *A Forest Atlas of the Northeast* (Upper Darby, PA, Northeastern Forest Experiment Station, 1968), pp. 23, 33; John Smith, *The Generall Historie of Virginia* (1624), facsimile reprint, Readex Microprint (1966), pp. 214-15. Note Verrazzano's poor identification of species; and note too that even Smith thought the forbidding Maine coastal forest might be obscuring more fertile country inland.

14. Maps of New England vegetation zones can be found in Lull, *Forest Atlas,* p. 7; Marinus Westveld, *et al.,* "Natural Forest Vegetation Zones of New England," *Journal of Forestry,* 54 (1956), pp. 332-8; and A. W. Kuchler's "Potential Natural Vegetation of the Contermi-nous United States," *American Geographical Society Special Publications,* 36 (1964), of which the most convenient copy is the one in the *National Atlas.* For a critique of such maps, see Margaret B. Davis, "Phytogeography and Palynology of Northeastern United States," in H. E. Wright and David G. Frey, eds., *The Quaternary of the United States* (Princeton, 1965), p. 381.

15. Rosier, "Voyage of George Waymouth," in Burrage, *Early Voyages,* pp. 384-5. Burrage believed that the river from which Rosier marched was the St. George's.

16. Higginson, *Plantation,* pp. 307-8.

17. Wood, pp. 38-9; John C. Huden, "Indian Place Names of New England," *Contributions from the Museum of the American Indian,* Heye Foundation, New York, 18 (1962)—see, for instance, entries for

Ohomowauke Swamp in Rhode Island and for Copecut, Massachusetts. Cf. Roger Williams, *A Key into the Language of America* (1643), John J. Teunissen and Evelyn J. Hinz, eds. (Detroit, 1973), p. 150.

18. Huden, "Indian Place Names," entries for Quawawehunk and Tatomuck; Morton, *Canaan*, pp 184-5. I have avoided cluttering the text with alternative interpretations, but should note that I may be misreading Morton here. He makes an explicit distinction between cedar and cypress which sounds to me like the one between red and white cedars. I have chosen to read his "cedar" as "white cedar" because doing so creates no real distortions in my characterization of the tree's habitat, but he may well have meant something else. There are no true cedars in North America, so the problem of nomenclature here is a vexing one.

19. Lull, *Forest Atlas*, pp. 19, 33; Smith, *Generall Historie*, p. 215; Alexander Young, ed., *Chronicles of the Pilgrim Fathers* (Boston, 1841), pp. 123-4.

20. Neil Jorgensen, *A Sierra Club Naturalist's Guide to Southern New England* (San Francisco, 1978), pp. 238-48; Silas Little, "Effects of Fire on Temperate Forests: Northeastern United States," in T. T. Kozlowski and C. E. Ahlgren, eds., *Fire and Ecosystems* (New York, 1974), esp. pp. 237-42.

21. Morton, *Canaan*, p. 184; Timothy Dwight, *Travels in New England and New York* (1821), Barbara Miller Solomon, ed. (Cambridge, MA, 1969), I, p. 21; IV, p. 151; Stanley W. Bromley, "The Original Forest Types of Southern New England," *Ecological Monographs*, 5 (1935), pp. 72-6. Note the nomenclature problem again: Morton could have been describing either a pitch pine forest or a tract of white pine on an outwash plain. The general argument I make here is nevertheless sound. For further details on fire ecology, see the discussion in Chapter 3.

22. Morton, *Canaan*, p. 227; Wood, *Prospect*, pp. 56-7; Emerson, *Letters*, p. 106.

23. Wood, *Prospect*, pp. 34, 51-2; John Smith, "Advertisements for the Unexperienced Planters of New-England," *Massachusetts Historical Society Collections*, 3rd series, 3 (1833), p. 37; Higginson, *Plantation*, p. 308; Emerson, *Letters*, p. 214. On the ecology of salt marshes, see Alfred C. Redfield, "Development of a New England Salt Marsh," *Ecological Monographs*, 42 (1972), pp. 201-37; and John and Mildred Teal's beautifully written *Life and Death of the Salt Marsh* (New York, 1969), esp. Part I.

24. On slope diversity and forest types, see Jorgensen, *Sierra Club Guide*, pp. 122-203, esp. 126-7; Bromley, "Original Forest Types"; George E. Nichols, "The Vegetation of Connecticut: II. Virgin Forests," *Torreya*, 13 (1913), pp. 199-215; and on northern hardwood forests, which I ignore here, A. C. Cline and S. H. Spurr, "The Virgin Upland Forest of Central New England," *Harvard Forest Bulletin*, 21 (1942), pp. 1-58. On the role of historical catastrophes in determining forest composition, see J. D. Henry and J. M. A. Swan, "Reconstructing

Forest History from Live and Dead Plant Material," *Ecology,* 55 (1974), pp. 772-83; C. D. Oliver and E. P. Stephens, "Reconstruction of a Mixed-Species Forest in Central New England," *Ecology,* 58 (1977), pp. 562-72, which is based on the data in E. P. Stephens, *The Historical-Developmental Method of Determining Forest Trends,* Ph.D. Thesis, Harvard University, 1955; and F. Herbert Bormann and Gene E. Likens, *Pattern and Process in a Forested Ecosystem* (New York, 1979). The technical debate here is whether forest succession is principally *autogenic* (driven by continuous processes internal to the forest ecosystem) or *allogenic* (driven by essentially stochastic processes external to the forest, such as fires or storms).

25. This paragraph is based on my reading of the fossil pollen studies which I describe in the bibliographical essay. On hemlock destruction, see Gene E. Likens and Margaret B. Davis, "Post-Glacial History of Mirror Lake and Its Watershed in New Hampshire," *Verhandlungen der Internationale Vereinigung für Theoretische und Angewandte Limnologie,* 19 (1975), p. 989; and Davis, "Phytogeography and Palynology," p. 394.

26. Morton, *Canaan,* pp. 175, 180.

3. Seasons of Want and Plenty

1. Christopher Levett, "Voyage into New England" (1628), *Massachusetts Historical Society Collections,* 3rd ser., 3 (1843), p. 179. Virginia probably suffered most from this kind of exaggeration, but see Thomas Morton, *New English Canaan* (1632), Charles F. Adams, ed., *Pubs. of the Prince Society,* XIV (Boston, 1883), pp. 231-3, for tendencies in the same direction. William Morrell's "Poem on New-England" (ca. 1623), *Massachusetts Historical Society Collections,* 1st ser., 1 (1792), pp. 125-39, is also a good example.

2. Quoted by Thomas Hutchinson, *The History of the Colony and Province of Massachusetts-Bay* (1765), Lawrence Shaw Mayo, ed. (Cambridge, MA, 1936), I, p. 405.

3. Levett, "Voyage," p. 182; *A Relation of the English Plantation at Plimoth* (1622), facsimile edition, Readex Microprint (1966) (henceforth cited as *Mourt's Relation*), p. 63; John [?] Pond to William Pond, March 15, 1630/1, in Everett Emerson, ed., *Letters from New England* (Amherst, MA, 1976), p. 65. See Neil Salisbury, *Manitou and Providence* (New York, 1982), pp. 81-2, 111-18, for evidence that Indian refusal to trade also lay at the root of the Sagadahoc and Wessagusset failures. On differences between American and European climates, see Karen Ordahl Kupperman, "The Puzzle of American Climate in the Early Colonial Period," *American Historical Review,* 87 (1982), pp. 1262-89.

4. John Smith, *The Generall Historie of Virginia* (1624), facsimile edition, Readex Microprint (1966), pp. 211, 219; Morrell, "New England," p.

131; Morton, *Canaan*, p. 177; James A. Henretta, *The Evolution of American Society, 1700-1815* (Lexington, MA, 1973), pp. 31-9. Exaggerated expectations were not limited to the English. Pierre Biard's *Relation* (in Reuben Gold Thwaites, ed., *Jesuit Relations, III, Acadia, 1611-1616* [Cleveland, 1897], pp. 65-7) supplies a wonderful French example: "... we Frenchmen are so willing to go there with our eyes shut and our heads down; believing, for example, that in Canada, when we are hungry, all we will have to do is go to an Island, and there by the skillful use of a club, right and left, we can bring down birds each as big as a duck, with every blow. This is well said, as our people have done this more than once and in more than one place. It is all very well, if you are never hungry except when these birds are on the Islands, and if even then you happen to be near them. But if you are fifty or sixty leagues away, what are you going to do?" Good discussions of these colonial assumptions about the New World can be found in Karen O. Kupperman, *Settling with the Indians* (Totowa, NJ, 1980); in Edmund S. Morgan's classic, "The Labor Problem at Jamestown, 1607-18," *American Historical Review*, 76 (1971), pp. 595-611; and his *American Slavery, American Freedom* (New York, 1975).

5. For general introductions to photoperiodism, see Robert L. Smith, *Ecology and Field Biology* (New York, 1966), pp. 98-126; Robert E. Ricklefs, *Ecology*, 2nd ed. (New York, 1979), pp. 280-306; and Edward J. Kormondy, *Concepts of Ecology* (Englewood Cliffs, NJ, 1969), pp. 140-54.

6. Roger Williams, *A Key into the Language of America* (1643), John J. Teunissen and Evelyn J. Hinz, eds. (Detroit, 1973), pp. 127-8; Morton, *Canaan*, p. 177. On the changing shape of wigwams, see William Wood, *New England's Prospect* (1634), Alden T. Vaughan, ed. (Amherst, MA, 1977), p. 113; H. P. Biggar, ed., *The Works of Samuel de Champlain*, 6 vols. (Toronto, 1922-36), maps; Daniel Gookin, "Historical Collections of the Indians in New England," *Massachusetts Historical Society Collections*, 1st ser., 1 (1792), p. 150; William C. Sturtevant, "Two 1761 Wigwams at Niantic, Connecticut," *American Antiquity*, 40 (1975), pp. 437-44; and Bernard G. Hoffman, *The Historical Ethnography of the Micmac in the Sixteenth and Seventeenth Centuries*, Ph.D. Thesis, UCLA, 1955, p. 135.

7. Lawrence C. Wroth, ed., *The Voyages of Giovanni de Verrazzano, 1524-1528* (New Haven, 1970), p. 140.

8. Hoffman, *Micmac Ethnography*, is superb on northern subsistence cycles; his diagram of these has been published in "Ancient Tribes Revisited," *Ethnohistory*, 14 (1967), p. 21. The northern documents of the French Jesuits are exceptionally fine: these include Biard, *Relation*, pp. 79-85; Nicolas Denys, *The Description and Natural History of the Coasts of North America (Acadia)* (1672), William F. Ganong, ed. (Toronto, Champlain Society Publications, II, 1908); and Chrestien Le Clercq, *New Relations of Gaspesia* (1691), William F. Ganong, ed.,

(Toronto, Champlain Society Publications, V, 1910). I have used them extensively in the discussion which follows, even though they fall outside the regional boundaries of New England, because documentary coverage of the Maine Indians is poor and northern New England Indians were much more like their Canadian neighbors than the Indians to the south. The best modern account of the Maine Indians is Frank G. Speck, *Penobscot Man* (Philadelphia, 1940). Quotations in this paragraph are from LeClercq, *Gaspesia*, p. 137; and Biard, *Relation*, p. 81. Hoffman, *Micmac Ethnography*, p. 160, argues that cod was not a major component of the northern coastal diet, despite Biard's claims to the contrary.

9. Biard, *Relation*, pp. 83, 101-3; Denys, *Acadia*, pp. 405, 422-3; Bruce J. Bourque, "Aboriginal Settlement and Subsistence on the Maine Coast," *Man in the Northeast*, 6 (Fall 1973), pp. 3-20; John Gyles, "Memoirs of Odd Adventures, Strange Deliverances, etc." (1736), in Alden T. Vaughan and Edward W. Clark, eds., *Puritans among the Indians* (Cambridge, MA, 1981), p. 103. Note again the nomenclature problem here: there are no bustards (Biard's *outardes)* in North America. We cannot be sure to which species he was referring.

10. Le Clercq, *Gaspesia*, p. 110; James Sullivan, "The History of the Penobscott Indians," *Massachusetts Historical Society Collections*, 1st ser., 9 (1804), p. 228; Biard, *Relation*, p. 107.

11. Eugene P. Odum, *Fundamentals of Ecology*, 3rd ed. (Philadelphia, 1971), pp. 106-39, explicates Liebig's Law and other environmental constraints on populations. It is unclear whether the starvation periods which French Jesuits observed among northern Indian populations at the beginning of the seventeenth century were typical of precolonial times. There is at least some reason to believe that the famines may have been the result of Indians having shifted their subsistence patterns to include trade with Europeans along the coast of Maine and Nova Scotia. Hoffman, *Micmac Ethnography*, pp. 229-33, gives the arguments against viewing the seventeenth-century starvations as normal; see also Bourque, "Aboriginal Settlement." On general hunter-gatherer behavior, see Richard B. Lee and Irven DeVore, eds., *Man the Hunter* (New York, 1968); and Marshall Sahlins, *Stone Age Economics* (Chicago, 1972).

12. M. K. Bennett, "The Food Economy of the New England Indians, 1605-75," *Journal of Political Economy*, 63 (1955), pp. 391-3. (Historians generally use Bennett's figures fairly uncritically, but there are many problems with them. See notes 16 and 20 below.)

13. The debate over pre-Columbian Indian population figures has generated an extensive literature, and my text conveys only the roughest outlines of its conclusions. See the bibliographical essay for a survey of this material. My own argument follows the discussions in Francis Jennings, *The Invasion of America* (Chapel Hill, 1975), pp. 15-31; S. F. Cook, *The Indian Populations of New England in the Seventeenth Century* (Berkeley, 1976); and especially Dean R. Snow, *The*

Archaeology of New England (New York, 1980), pp 31-42, whose density figures I have converted from square kilometers to square miles. Colonial population figures are from U. S. Bureau of the Census, *Historical Statistics of the United States* (Washington, 1975), Table Z 1-19, p. 1168.

14. Northern month names can be found in Le Clercq, *Gaspesia*, pp. 137-9; Hoffman, *Micmac Ethnography*, p. 246; Biard, *Relation*, pp. 79-83; and Philip K. Bock, "Micmac," in Trigger, *Northeast*, p. 111. Southern calendars are in Eva L. Butler, "Algonkian Culture and Use of Maize in Southern New England," *Bulletin of the Archaeological Society of Connecticut*, 22 (December 1948), pp. 10-11; "Indian Names of the Months," *New England Historical and Genealogical Register*, 10 (1856), p. 166; and Gordon M. Day, "An Agawam Fragment," *International Journal of American Linguistics*, 33 (1967), pp. 244-7. See also the discussion in Peter A. Thomas, "Contrastive Subsistence Strategies and Land Use as Factors for Understanding Indian-White Relations in New England," *Ethnohistory*, 23 (1976), pp. 1-18. The Abenaki calendar given by Sebastian Rasles in *Dictionary of the Abnaki Language in North America* (Cambridge, MA, 1833), p. 478, includes two or three months referring to maize cultivation.

15. Isaack de Rasieres to Samuel Blommaert, ca. 1628, in Sydney V. James, Jr., ed., *Three Visitors to Early Plymouth* (Plimoth Plantation, 1963), p. 71; Samuel de Champlain, *Voyages of Samuel de Champlain*, W. L. Grant, ed. (New York, 1907), p. 62. On multiple crop farming, see Fulmer Mood, "John Winthrop, Jr., on Indian Corn," *New England Quarterly*, 10 (1937), pp. 128-9; Williams, *Key*, pp. 170-1; Carl Sauer, "The Agency of Man on the Earth," in William L. Thomas, ed., *Man's Role in Changing the Face of the Earth* (Chicago, 1956), pp. 56-7; Harold C. Conklin, "An Ethnoecological Approach to Shifting Agriculture," *Transactions of the New York Academy of Science*, 2nd ser., 17 (1954), pp. 133-42; Conklin, "The Study of Shifting Cultivation," *Current Anthropology*, 2 (1961), pp. 27-61; and Howard S. Russell, "New England Indian Agriculture," *Bulletin of the Massachusetts Archaeological Society*, 22 (1961), pp. 58-61.

16. Williams, *Key*, p. 121; Judith K. Brown, "A Note on the Division of Labor by Sex," *American Anthropologist*, 72 (1970), pp. 1073-8. Note that I refer here only to the sexual division of food-producing activities; I make no effort to consider the total allocation of physical or non-physical work. On the productivity of maize agriculture, see Peter A. Thomas, *In the Maelstrom of Change: The Indian Trade and Cultural Process in the Middle Connecticut River Valley, 1635-1665*, Ph.D. Thesis, University of Massachusetts, 1979, p. 109; Williams, *Key*, p. 171; Bennett, "Food Economy," pp. 391-3. Bennett's figures for corn's contribution to the Indian diet are probably exaggerated: he derived them by estimating total caloric requirements for an average person, subtracting Williams's corn yield estimates as distributed on a per capita basis, and allocating the remainder to noncorn foods. Such

an algorithm obviously privileges corn at the expense of other foods, fails to consider waste, and assumes that corn was consumed at a constant level all year long. My discussion in the text shows why I think this unlikely; see also note 20 below.

17. Mood, "Winthrop on Corn," p. 126; Wood, *Prospect*, p. 35; James, *Plymouth Visitors*, pp. 7-9. The case against Indian fish fertilizer was first made by Erhard Rostlund, "The Evidence for the Use of Fish as Fertilizer in Aboriginal North America," *Journal of Geography*, 56 (1957), pp. 222-8; and Lynn Ceci has put forward the strongest collection of arguments in "Fish Fertilizer: A Native North American Practice?" *Science*, 188 (1975), pp. 26-30. She replies to critics in *Science*, 189 (1975), pp. 946-50. On the exhaustion of grain stores by late winter, see Lorraine E. Williams, *Ft. Shantok and Ft. Corchaug: A Comparative Study of Seventeenth Century Culture Contact in the Long Island Sound Area*, Ph.D. Thesis, New York University, 1972, p. 232, which confirms my critique of Bennett with the help of archaeological evidence.

18. James, *Plymouth Visitors*, p. 79; Williams, *Key*, pp. 128, 163; Wood, *Prospect*, p. 114; Butler, "Maize," pp. 18-19; Mood, "Winthrop on Corn," p. 126; John Josselyn, "An Account of Two Voyages to New-England" (1675), *Massachusetts Historical Society Collections*, 3rd ser., 3 (1833), p. 296.

19. Williams, *Key*, pp. 176, 178; Denys, *Acadia*, pp. 407, 420-2; Biard, *Relation*, p. 83; Josselyn, *Two Voyages*, pp. 305-6.

20. Butler, "Maize," pp. 24-6; Williams, *Key*, p. 231; Wood, *Prospect*, pp. 103-5; Gookin, "Historical Collections," p. 153. Thomas, *Maelstrom of Change*, p. 348, gives archaeological evidence that there was a greater fall consumption of berries than historical sources note. On the range of foods gathered, see Lucia S. Chamberlain, "Plants Used by the Indians of Eastern North America," *American Naturalist*, 35 (1901), pp. 1-10; Gretchen Beardsley, "The Groundnut as Used by the Indians of Eastern North America," *Papers of the Michigan Academy of Sciences, Arts, and Letters*, 25 (1940), pp. 507-15; and Frederic W. Warner, "The Foods of the Connecticut Indians," *Bulletin of the Archaeological Society of Connecticut*, 37 (1972), pp. 27-47. The massive fall consumption of corn in these festivals argues against the assumption in Bennett's overly aggregated statistics that corn was eaten at a constant level year round.

21. Williams, *Key*, pp. 128, 224; Wood, *Prospect*, p. 106; *Mourt's Relation*, p. 8; Thomas, *Maelstrom of Change*, pp. 106-8.

22. Josselyn, *Two Voyages*, pp. 302-4; Morton, *Canaan*, p. 138; Butler, "Maize," pp. 28-30; Gookin, "Historical Collections," p. 150; George Lyman Kittredge, ed., "Letters of Samuel Lee and Samuel Sewall Relating to New England and the Indians," *Publications of the Colonial Society of Massachusetts*, 14 (1912), p. 148.

23. Thomas, *Maelstrom of Change*, pp. 355-7; Josselyn, *Two Voyages*, p. 297. My arguments about the relationship between hunting, meat, and

clothing are derived from Harold Hickerson, "The Virginia Deer and Intertribal Buffer Zones in the Upper Mississippi Valley," in Anthony Leeds and Andrew P. Vayda, eds., *Men, Culture and Animals*, AAAS, Publication No. 78 (1965), pp. 43-65; and Richard Michael Gramly, "Deerskins and Hunting Territories: Competition for a Scarce Resource of the Northeastern Woodlands," *American Antiquity*, 42 (1977), pp. 601-5. Gramly overstates the importance of deer, but his arguments are otherwise sound.

24. "Extract from an Indian History," *Massachusetts Historical Society Collections*, 1st ser., 9 (1804), p. 101; a paraphrased version of this can be found in Hedrick Aupaumut, *First Annual Report of the American Society for Promoting the Civilization and General Improvement of the Indian Tribes of the United States* (New Haven, 1824), pp. 41-2.

25. Morton, *Canaan*, p. 138; Williams, *Key*, pp. 107, 138; see also William Christie MacLeod, "Fuel and Early Civilization," *American Anthropologist*, N.S., 27 (1925), pp. 344-6; Wroth, *Verrazzano*, p. 139; Higginson, *Plantation*, p. 308. Le Clerq gives us a nice portrait of the garbage problems which also led Indians to move their campsites: "They are filthy and vile in their wigwams, of which the approaches are filled with excrements, feathers, chips, shreds of skins, and very often with entrails of the animals or the fishes which they take in hunting or fishing" *(Gaspesia*, p. 253).

26. Quotations are from Morton, *Canaan*, p. 172; Wood, *Prospect*, p. 38; and, on fending off invaders, Martin Pring, "A Voyage Set Out from the Citie of Bristoll, 1603," in Henry S. Burrage, ed., *Early English and French Voyages* (New York, 1906), p. 351. Other primary documents on burning are Higginson, p. 308; Edward Johnson, *Johnson's Wonder-Working Providence*, J. Franklin Jameson, ed. (New York, 1910), p. 85; Benjamin Trumbull, *A Complete History of Connecticut* (Hartford, 1797), p. 23. The classic essay on Indian burning is Gordon M. Day, "The Indian as an Ecological Factor in the Northeastern Forest," *Ecology*, 34 (1953), pp. 329-46; see also the various articles cited in the bibliographical essay. The chief critique of this interpretation of Indian burning is Hugh M. Raup, "Recent Changes of Climate and Vegetation in Southern New England and Adjacent New York," *Journal of the Arnold Arboretum*, 18 (1937), pp. 79-117. Raup believed that "to picture a wholesale conflagration in Massachusetts, Rhode Island, Connecticut, and southern New York State as would involve most of the inflammable woods every year, or even every 10 to 20 years, is inconceivable" (p. 84). In this, he failed to take account of the reduced fuel burden at ground level in forests which are repeatedly burned; the very fact of regular burning kept ground fires from reaching the canopy. But Raup was no doubt right that the *entirety* of southern New England was never regularly burned; I have limited the claims of my argument to the local vicinity of village sites. A recent article defends Raup but basically confirms my emphasis on local burning: Emily W. B.

Russell, "Indian-Set Fires in the Forests of the Northeastern United States," *Ecology,* 64 (1983), pp. 78–88.

27. Regina Flannery, "An Analysis of Coastal Algonquian Culture," *Catholic University Anthropological Series,* 7 (1939), pp. 14, 167; Day, "Indian as Ecological Factor," pp 338-9. Northern hardwood forests in the Midwest were probably burned completely by lightning and man-made fires on a hundred-year cycle, but the same was apparently not true of the northern New England forest. See Orrie L. Loucks, "Evolution of Diversity, Efficiency and Community Stability," *American Zoologist,* 10 (1970), pp. 17-25; Sidney S. Frissell, Jr., "The Importance of Fire as a Natural Ecological Factor in Itasca State Park, Minnesota," *Quaternary Research,* 3 (1973), pp. 397-407; Craig C. Lorimer, "The Presettlement Forest and Natural Disturbance Cycle of Northeastern Maine," *Ecology,* 58 (1977), pp. 139-48; and F. Herbert Bormann and Gene E. Likens, "Catastrophic Disturbance and the Steady State in Northern Hardwood Forests," *American Scientist,* 67 (1979), pp. 660-9.

28. Williams, *Key,* pp. 165, 168, 191. On fire effects discussed in this and the subsequent paragraph, see William A. Niering, *et al.,* "Prescribed Burning in Southern New England: Introduction to Long-Range Studies," *Proceedings of the Annual Tall Timbers Fire Ecology Conference,* 10 (1970), pp. 267-86; Silas Little, "Effects of Fire on Temperate Forests: Northeastern United States," in T. T. Kozlowski and C. E. Ahlgren, eds., *Fire and Ecosystems* (New York, 1974), pp. 225-50; and the articles in the special October 1973 issue of *Quaternary Research,* especially the superb summary in the Introduction by H. E. Wright, Jr., and M. L. Heinselman, pp. 319-28.

29. Odum, *Fundamentals of Ecology,* pp. 157-9; Timothy Dwight, *Travels in New England and New York* (1821), Barbara Miller Solomon, ed. (Cambridge, MA, 1969), IV, pp. 38-9; Williams, *Key,* p. 165. E. L. Jones discusses these edge effects in his "Creative Disruptions in American Agriculture, 1620-1820," *Agricultural History,* 48 (1974), pp. 514-15, although I think he becomes confused when he describes burned forests as dense, dark, and deep, with few birds. Exactly the opposite was the case.

30. Quotations on women's work are from Williams, *Key,* p. 122 (in which he refers specifically to the loads women bear in moving camp); Levett, "Voyage," p. 178; and Thomas Lechford, *Plain Dealing* (1642), *Massachusetts Historical Society Collections,* 3rd ser., 3 (1833), p. 103. See also Morrell, "New England," p. 136; and Wood, *Prospect,* pp. 92, 115-16. Two recent articles assert a much more egalitarian relationship between Indian men and women: Robert Steven Grumet, "Sunksquaws, Shamans, and Tradeswomen: Middle Atlantic Coastal Algonkian Women During the 17th and 18th Centuries," in Mona Etienne and Eleanor Leacock, eds., *Women and Colonization* (New York, 1980), pp. 43-62; and Trudie Lamb, "Squaw Sachems: Women Who Rule," *Artifacts,* 9:2 (Winter/Spring 1981), pp. 1-3.

These both argue from the experience of elite women in leadership roles, and Grumet seems to me unsuccessful in proving his claim that the sexual division of labor was not fairly strict.

31. Francis Higginson, *New-Englands Plantation* (1630), *Massachusetts Historical Society Proceedings*, 62 (1929), p. 316. For further materials on sexual work roles, see James Axtell, ed., *The Indian Peoples of Eastern America: A Documentary History of the Sexes* (New York, 1981), pp. 103-39.

32. Wroth, *Verrazzano*, p. 139. In areas where a diversity of habitats was lacking, like central Vermont, Indian populations were very low or absent. See William A. Bayreuther, "Environmental Diversity as a Factor in Modeling Prehistoric Settlement Patterns: Southeastern Vermont's Black River Valley," *Man in the Northeast*, 19 (1980), pp. 83-93. But see also Gordon M. Day, "The Indian Occupation of Vermont," *Vermont History*, 33 (1965), pp. 365-74.

4. Bounding the Land

1. Thomas Morton, *New English Canaan* (1632), Charles F. Adams, ed., *Pubs. of the Prince Society*, XIV (Boston, 1883), pp. 175-7.

2. Francis Higginson, *New-Englands Plantation* (1630), in *Massachusetts Historical Society Proceedings*, 62 (1929), p. 316; William Wood, *New England's Prospect* (1634), Alden T. Vaughan, ed. (Amherst, 1977), p. 96; Izaak Walton, *The Compleat Angler* (London, 1653).

3. Robert Cushman, "Reasons and Considerations Touching the Lawfulness of Removing Out of England into the Parts of America" (1621), in Alexander Young, ed., *Chronicles of the Pilgrim Fathers* (Boston, 1841), p. 243; see also James Sullivan, *The History of Land Titles in Massachusetts* (Boston, 1801), pp. 21-27; Charles E. Eisinger, "The Puritans' Justification for Taking the Land," *Essex Institute Historical Collections*, 84 (1948), pp. 131-43; and Ruth Barnes Moynihan, "The Patent and the Indians: The Problem of Jurisdiction in Seventeenth-Century New England," *American Indian Culture and Research*, 2:1 (1977), pp. 8-18.

4. John Winthrop, "Reasons to Be Considered, and Objections with Answers," *Winthrop Papers*, Massachusetts Historical Society, II (1931), pp. 140-1; John Winthrop, *Winthrop's Journal*, James Kendall Hosmer, ed. (New York, 1908), p. 294; John Cotton, "Gods Promise to His Plantations," *Old South Leaflets* (nd), no. 53, p. 6. See also Francis Jennings, *The Invasion of America* (Chapel Hill, 1975), pp. 135-6.

5. John Cotton, "John Cotton's Answer to Roger Williams," in *The Complete Writings of Roger Williams* (New York, 1963), II, pp. 46-7. Williams of course had fully developed theological reasons for his position which I do not mean to discount; but these were not incon-

sistent with protecting land claims purchased from the Indians. For an analogous argument, see Jeremiah Dummer, *A Defence of the New England Charters* (London, 1721), pp. 8-12. Jennings, *Invasion,* esp. pp. 43-84, is excellent on European conquest ideologies; see also Fred M. Kimmey, "Christianity and Indian Lands," *Ethnohistory,* 7 (1960), pp. 44-60.

6. Cotton, "Answer," p. 47. A fine discussion of this subject which parallels my own in ecological emphasis is Peter A. Thomas, "Contrastive Subsistence Strategies and Land Use as Factors for Understanding Indian-White Relations in New England," *Ethnohistory,* 23 (1976), pp. 1-18.

7. Huntington Cairns, *Law and the Social Sciences* (New York, 1935), p. 59. The classic anthropological essay on property is A. Irving Hallowell, "The Nature and Function of Property as a Social Institution," *Journal of Legal and Political Sociology,* 1 (1942-3), pp. 115-38; see also E. Adamson Hoebel, *The Law of Primitive Man* (Cambridge, MA, 1967), pp. 46-63. For an economist's discussion of the same issues, one that seems to me to misconstrue the social and ecological nature of property rights, see Harold Demsetz, "Toward a Theory of Property Rights," *American Economic Review Papers and Proceedings,* 57 (1967), pp. 347-59.

8. John Josselyn, *An Account of Two Voyages to New-England* (1675), in *Massachusetts Historical Society Collections,* 3rd ser., 3 (1833), p. 308; Roger Williams, *A Key into the Language of America* (1643), John J. Teunissen and Evelyn J. Hinz, eds. (Detroit, 1973), p. 201; Wood, *Prospect,* pp. 97-8. See also Daniel Gookin, "Historical Collections of the Indians in New England," *Massachusetts Historical Society Collections,* 1st ser., 1 (1792), p. 154; Chrestien Le Clercq, *New Relation of Gaspesia* (1691), William F. Ganong, ed. (Toronto, Champlain Society Publications, V, 1910), pp. 234-5; and John M. Cooper, "Is the Algonquian Family Hunting Ground System Pre-Columbian?," *American Anthropologist,* 41 (1939), p. 71. For a review of the literature on Indian property systems, see the bibliographical essay.

9. Gookin, "Collections," p. 154. References to historians who have sought to show similarities between Indian and European polities can be found in the bibliographical essay.

10. Edward Winslow, *Good Newes from New England* (1624), in Alexander Young, ed., *Chronicles of the Pilgrim Fathers* (Boston, 1841), p. 361; Williams, *Key,* p. 167; see also Pierre Biard, *Relation,* in Reuben Gold Thwaites, ed., *Jesuit Relations, III, Acadia, 1611-1616* (Cleveland, 1897), p. 89.

11. Williams, *Key,* p. 167; Williams, *The Letters of Roger Williams,* John Russell Bartlett, ed., *Pubs. of the Narragansett Club* (Providence, 1874), IV, p. 104.

12. Williams, *Key,* p. 152; William S. Simmons, *Cautantowwit's House* (Providence, 1970), pp. 45-6; Susan Gibson, ed., *Burr's Hill* (Providence, 1980), pp. 14-21; Melville J. Herskovits, *Economic Anthropology* (New York, 1952), pp. 371-92; Morton, *Canaan,* p. 177.

13. Le Clercq, *Gaspesia,* p. 245; Biard, *Relation,* p. 89; Anthony F. C.

Wallace, "Woman, Land, and Society: Three Aspects of Aboriginal Delaware Life," *Pennsylvania Archaeologist*, 17 (1947), p. 2.

14. John M. Cooper, "Land Tenure among the Indians of Eastern and Northern North America," *Pennsylvania Archaeologist*, 8 (1938), pp. 55-9; Froehlich G. Rainey, "A Compilation of Historical Data Contributing to the Ethnography of Connecticut and Southern New England Indians," *Bulletin of the Archaeological Society of Connecticut*, 3 (April 1936), p. 60; Nathaniel B. Shurtleff, ed., *Records of the Governor and Company of Massachusetts Bay in New England* (Boston, 1853), III, p. 281. Iroquois Indian women specifically owned their wigwams and planting fields; New England Indian women may well have had similar rights, but there is no direct evidence to corroborate this.

15. Cooper, "Land Tenure," p. 56; Eleanor Leacock, "The Montagnais 'Hunting Territory' and the Fur Trade," *Memoirs of the American Anthropological Association*, 78 (1954), p. 2; Morton, *Canaan*, p. 138.

16. Morton, *Canaan*, p. 162; Williams, *Key*, p. 224.

17. Williams, *Key*, p. 224. One of the best documented instances of how an animal's habits affects the way it is hunted and owned is the difference between moose and caribou in the subarctic North. Moose are solitary, while caribou move in herds; one is hunted by small bands with well-defined territorial rights, the other by large groups with relatively unfixed territorial boundaries. See René R. Gadacz, "Montagnais Hunting Dynamics in Historicoecological Perspective," *Anthropologica*, 17 (1975), pp. 149-67; and Rolf Knight, "A Re-examination of Hunting, Trapping, and Territoriality among the Northeastern Algonkian Indians," in Anthony Leeds and Andrew P. Vayda, eds., *Man, Culture and Animals*, AAAS Publication No. 78 (1965), pp. 27-42.

18. Le Clercq, *Gaspesia*, p. 263; Regina Flannery, *An Analysis of Coastal Algonquian Culture*, Catholic U. Anthropological Series, 7 (1939), p. 78. On the family hunting territory debate, see Chapter 5.

19. Herskovits, *Economic Anthropology*, pp. 331-70; Marshall Sahlins, *Tribesmen* (Englewood Cliffs, NJ, 1968), p. 76; John C. Huden, "Indian Place Names of New England," *Contributions from the Museum of the American Indian*, Heye Foundation, New York, 18 (1962); Williams, *Key*, p. 171; Eva L. Butler, "Algonkian Culture and Use of Maize in Southern New England," *Bulletin of the Archaeological Society of Connecticut*, 22 (1948), pp. 27-8.

20. Huden, "Place Names."

21. *Ibid.* For examples of boundary place-names, see Acqueet, Atchaubennuck, Bunganut, Chaubongum, Neutaqunkanet, etc.

22. Harry Andrew White, ed., *Indian Deeds of Hampden County* (Springfield, MA, 1905), pp. 11-12. Although this deed was not actually recorded until 1679, its construction suggests that it genuinely dated from 1636 as it claimed to do. For a discussion of Agawam land transactions generally, see Peter A. Thomas, *In the Maelstrom of Change*, Ph.D. Thesis, University of Massachusetts, 1979, pp. 133-44.

23. White, *Indian Deeds*, p. 12.

24. Jennings, *Invasion*, p. 129; Shurtleff, *Massachusetts Records*, I, p. 400.

The company also assured its agents that, if they were careful, "the natives wilbe willing to treat and compound with you upon very easie conditions."

25. Sidney Perley, ed., *The Indian Land Titles of Essex County, Massachusetts* (Salem, MA, 1912), p. 26.

26. Wilcomb E. Washburn, "The Moral and Legal Justifications for Dispossessing the Indians," in James Morton Smith, ed., *Seventeenth-Century America* (Chapel Hill, 1959), pp. 15-32; Marshall Harris, *Origin of the Land Tenure System of the United States* (Ames, IA, 1953), pp. 155-78.

27. John Bulkley, "An Inquiry into the Right of the Aboriginal Natives to the Lands in America . . . ," *Massachusetts Historical Society Collections*, 1st ser., 4 (1795), pp. 172-3; Shurtleff, *Massachusetts Records*, I, p. 112; *Laws of the Colonial and State Governments Relating to the Indians and Indian Affairs* (Washington, 1832), pp. 10, 40-1, 53, 59; Gookin, "Collections," p. 179; Charles J. Hoadly, ed., *The Public Records of the Colony of Connecticut, 1717-1725* (Hartford, 1872), p. 13. See also Roy H. Akagi, *The Town Proprietors of the New England Colonies* (Philadelphia, 1924), pp. 22-30; Yasu Kawashima, "Legal Origins of the Indian Reservation in Colonial Massachusetts," *American Journal of Legal History*, 13 (1969), pp. 42-56; and the suggestive discussion of a later period in John T. Juricek, "American Usage of the Word 'Frontier' from Colonial Times to Frederick Jackson Turner," *Proceedings of the American Philosophical Society*, 110 (1966), pp. 10-34.

28. Melville Egleston, "The Land System of the New England Colonies," *Johns Hopkins University Studies in Historical and Political Science*, 4th ser., 11-12 (1886), pp. 6-7; Shurtleff, *Massachusetts Records*, I, p. 5; Harris, *Land Tenure System*, pp. 37-9; Viola Florence Barnes, "Land Tenure in English Colonial Charters of the Seventeenth Century," in *Essays in Colonial History Presented to Charles McLean Andrews by His Students* (New Haven, 1931), pp. 13-16.

29. Shurtleff, *Massachusetts Records*, I, pp. 3-4.

30. Egleston, "New England Land System," p. 19; Shurtleff, *Massachusetts Records*, I, pp. 363-4, 399. Technically, the Crown did not in fact grant to the Massachusetts Bay Company the right to create self-governing communities with delegated political authority, but since the Company behaved as if it *did* have this right, I have ignored this issue in my discussion in the text.

31. "Essay on the Ordering of Towns," *Winthrop Papers* (Boston, 1943), III, p. 183. On functional divisions of land, see Herbert L. Osgood, *The American Colonies in the Seventeenth Century* (New York, 1904), I, pp. 437-40; Charles M. Andrews, "The River Towns of Connecticut," *Johns Hopkins University Studies in Historical and Political Science*, 7th ser., 7-9 (1889), pp. 32-81; Percy Wells Bidwell and John I. Falconer, *History of Agriculture in the Northern United States, 1620-1860* (Washington, D.C., 1925), pp. 49-60; Akagi, *Town Proprietors*, pp. 103-14; Harris, *Land Tenure System*, pp. 273-88; and the various town studies cited in the bibliographical essay.

32. Shurtleff, *Massachusetts Records*, I, pp. 90, 114; Leonard W. Labaree,

Milford, Connecticut: The Early Development of a Town as Shown in Its Land Records (New Haven, 1933), p. 15; David Grayson Allen, *In English Ways* (Chapel Hill, 1981), *passim;* "Essay on Ordering Towns," p. 184.

33. Allen, *English Ways;* Sumner Chilton Powell, *Puritan Village* (Middletown, CT, 1963); John J. Waters, "The Traditional World of the New England Peasants: A View from Seventeenth-Century Barnstable," *New England Historical and Genealogical Register,* 130 (1976), pp. 3-21; T. H. Breen, "Persistent Localism: English Social Change and the Shaping of New England Insitutions," *William and Mary Quarterly,* 3rd ser., 32 (1975), pp. 3-28; Breen, "Transfer of Culture: Chance and Design in Shaping Massachusetts Bay, 1630-1660," *New England Historical and Genealogical Register,* 132 (1978), pp. 3-17; Joan Thirsk, "The Farming Regions of England," in Joan Thirsk, ed., *The Agrarian History of England and Wales, IV, 1500-1640* (Cambridge, 1967), pp. 1-112.

34. White, *Indian Deeds,* p. 180. On the rise of land markets and the development of recording systems, see the bibliographical essay.

35. On shifting market and property conceptions, see Richard Schlatter, *Private Property* (New Brunswick, NJ, 1951); J. E. Crowley, *This Sheba, Self: The Conceptualization of Economic Life in Eighteenth-Century America* (Baltimore, 1974); Joyce Oldham Appleby, *Economic Thought and Ideology in Seventeenth-Century England* (Princeton, 1978). On market regulation in New England itself, suggestive discussions are Jon C. Teaford, *The Municipal Revolution in America* (Chicago, 1975); Gary Nash, *The Urban Crucible* (Cambridge, MA, 1979). Portrayals of New England as a subsistence economy are many; I have cited the more prominent examples (and counter-examples) in the bibliographical essay. I differ from these descriptions of the New England economy not in their specific details, which seem to me to be generally accurate, but in my sense that "subsistence" is an extremely relative concept; I am also inclined to see the existence of an alienated market in land as a crucial element in the transition to capitalism, one nearly as important to capital accumulation as alienated markets in labor.

36. *Oxford English Dictionary.*

37. The ability to think of land in terms of abstract equivalencies did not require actual market exchange. Labaree, *Early Development of Milford,* p. 5, describes how initial divisions of land in that town and others were made by defining a single tract of average size and quality that would serve as a standard "pattern" for other lots. Divisions were then made by "sizers," who gave each town proprietor a tract larger or smaller than the standard, depending on how they evaluated its relative quality.

38. *New England's First Fruits* (1643), in *Massachusetts Historical Society Collections,* 1st ser., 1 (1792), p. 250; Thomas Cooper, *Some Information Respecting America* (Dublin, 1794) (reprinted New York, 1969), p. 2.

39. John Locke, *Two Treatises of Government,* Peter Laslett, ed., rev. ed. (New York, 1963), p. 343. For Locke's confused notion of capital, see

C. B. Macpherson, "Locke on Capitalist Appropriation," *Western Political Quarterly*, 4 (1951), pp. 550-66, which my argument follows only in part; Eric Roll, *A History of Economic Thought*, 4th ed. (London, 1973), pp. 112-17; and Laslett's introduction to Locke, *Two Treatises*, pp. 111-20.

40. Locke, *Two Treatises*, pp. 338-9.

41. Marshall Sahlins, *Stone Age Economics* (Chicago, 1972), pp. 1-2; Morton, *Canaan*, p. 177; Biard, *Relation*, pp. 85, 135; cf. Nicolas Denys, *The Description and Natural History of the Coasts of North America (Acadia)* (1672), William F. Ganong, ed. (Toronto, Champlain Society Pubs., II, 1908), p. 426. Anyone familiar with Sahlins's work will recognize the debt that this entire chapter owes to it. See also Richard B. Lee, "What Hunters Do for a Living, or, How to Make Out on Scarce Resources," in Richard B. Lee and Irven DeVore, *Man the Hunter* (New York, 1968), pp. 30-48. Morton has been underrated as a colonial source for too long; he was a man of strongly held opinions—opinions acceptable neither to the Puritans nor to their filiopietistic descendants—but Samuel Maverick described *New English Canaan* as "the truest discription of New England as then it was that ever I saw." Samuel Maverick, "A Briefe Discription of New England," *Massachusetts Historical Society Proceedings*, 2nd ser., 1 (1884-85), p. 238.

42. Timothy Dwight, *Travels in New England and New York* (1821), Barbara Miller Solomon, ed. (Cambridge, MA, 1969), III, p. 18.

5. Commodities of the Hunt

1. Dean R. Snow, "Abenaki Fur Trade in the Sixteenth Century," *Western Canadian Journal of Anthropology*, 6 (1976), pp. 3-11; William I. Roberts III, *The Fur Trade of New England in the Seventeenth Century*, Ph.D. Thesis, University of Pennsylvania, 1958, pp. 1-18; Elspeth M. Veale, *The English Fur Trade in the Later Middle Ages* (Oxford, 1966).

2. Lawrence C. Wroth, ed., *The Voyages of Giovanni de Verrazzano, 1524-1528* (New Haven, 1970), pp. 138, 140.

3. John Brereton, "Briefe and True Relation of . . . Virginia" (1602), in Henry S. Burrage, ed., *Early English and French Voyages* (New York, 1906), pp. 337-8; Samuel de Champlain, *Voyages of Samuel de Champlain, 1604-1618*, W. L. Grant, ed. (New York, 1907), p. 50; Snow, "Abenaki Fur Trade," p. 6.

4. *Journall of the English Plantation at Plimoth*, hereafter cited as *Mourt's Relation* (1622), reprinted Readex Microprint (1966), p. 11. On burial practices, compare Pierre Biard, *Relation*, in Reuben Gold Thwaites, ed., *Jesuit Relations, III, Acadia, 1611-1616* (Cleveland, 1897), p. 131; and Roger Williams, *A Key into the Language of America* (1643), John J. Teunissen and Evelyn J. Hinz, eds. (Detroit, 1973), pp. 247-9.

On Europeans choosing to become Indians, see James Axtell, *The European and the Indian* (New York, 1981), pp. 131-206. The Pilgrims' sailor could have been one of the five Frenchmen whom Thomas Morton says were shipwrecked at Massachusetts Bay circa 1615, but there is no way to know. See Thomas Morton, *New English Canaan* (1632), Charles F. Adams, ed., *Pubs. of the Prince Society*, XIV (Boston, 1883), pp. 130-2.

5. On European diseases generally, see Alfred Crosby, *The Columbian Exchange* (Westport, CT, 1972); William H. McNeill, *Plagues and Peoples* (New York, 1976), p. 178; McNeill, *The Human Condition* (Princeton, 1980). The quotation is from William Wood, *New England's Prospect* (1634), Alden T. Vaughan, ed. (Amherst, MA, 1977), p. 110.

6. Alfred W. Crosby, "Virgin Soil Epidemics as a Factor in the Aboriginal Depopulation in America," *William and Mary Quarterly*, 3rd ser., 33 (1976), pp. 292-4.

7. Biard, *Relation*, p. 105.

8. Morton, *Canaan*, pp. 132-3. On the 1616-19 epidemic, see Charles Francis Adams, *Three Episodes of Massachusetts History* (1892) (reprinted New York, 1965), pp. 1-12; Sherburne F. Cook, "The Significance of Disease in the Extinction of the New England Indians," *Human Biology*, 45 (1973), pp. 485-508; Billee Hoornbeck, "An Investigation into the Cause or Causes of the Epidemic Which Decimated the Indian Population of New England, 1616-1619," *New Hampshire Archaeologist*, 19 (1976-77), pp. 35-46; William Bradford, *Of Plymouth Plantation*, Samuel Eliot Morison, ed. (New York, 1952), p. 87. Note that the northern limits of the epidemic apparently corresponded closely to those of maize agriculture. Why the disease did not extend to the Narragansetts is a mystery, but it may suggest the extent to which the two areas were isolated from each other in terms of trade. Indians told Roger Williams that epidemics had followed earthquakes in 1568, 1574, 1584, and 1592, but unlike Peter Thomas, I am inclined to be suspicious of the earthquake association and to reject this as evidence of earlier epidemics. See Roger Williams to John Winthrop [June 1638], *The Letters of Roger Williams*, John Russell Bartlett, ed. (Providence, 1874), p. 99; and Peter A. Thomas, *In the Maelstrom of Change: The Indian Trade and Cultural Process in the Middle Connecticut River Valley, 1635-1665*, Ph.D. Thesis, University of Massachusetts, 1979, p. 27.

9. Edward Winslow, *Good Newes from New England* (1624), in Alexander Young, ed., *Chronicles of the Pilgrim Fathers* (Boston, 1841), p. 305; Bradford, *Plymouth*, pp 270-1; John Duffy, "Smallpox and the Indians in the American Colonies," *Bulletin of the History of Medicine*, 25 (1951), pp. 324-41.

10. Winslow, *Good Newes*, p. 362; Williams, *Key*, p. 243; Bradford, *Plymouth*, p. 271.

11. Cook, "New England Indian Disease," pp. 501-3; Frances Jennings,

The Invasion of America (Chapel Hill, 1975), p. 29; Dean R. Snow, *The Archaeology of New England* (New York, 1980), p. 34.

12. Bradford, *Plymouth*, pp. 99, 271; *Mourt's Relation*, pp. 32-3; Winslow, *Good Newes*, pp. 289-92; Morton, *Canaan*, p. 245; Thomas, *Maelstrom*, p. 44. On maneuverings of Indians for power, compare the discussions of the Pequot leader Uncas in Jennings, *Invasion*, and P. Richard Metcalf, "Who Should Rule at Home: Native American Politics and Indian-White Relations," *Journal of American History*, 61 (1974), pp. 651-65. On disease and political destabilization, see Karl H. Schlesier, "Epidemics and Indian Middlemen: Rethinking the Wars of the Iroquois, 1609-1653," *Ethnohistory*, 23 (1976), pp. 129-45.

13. Edward Johnson, *Johnson's Wonder-Working Providence*, J. Franklin Jameson, ed. (New York, 1910), pp. 41, 80; John Winthrop, *Winthrop's Journal*, James K. Hosmer, ed. (New York, 1908), I, p. 115; Morton, *Canaan*, pp. 130-4; Robert Cushman, "Discourse," in Young, *Chronicles*, p. 258.

14. *Mourt's Relation*, p. 23; Howard S. Russell, *A Long, Deep Furrow* (Hanover, NH, 1976), p. 21; *New England's First Fruits* (1643), in *Massachusetts Historical Society Collections*, 1st ser., 1 (1792), p. 246; John Winthrop to Sir Simond D'Ewes, July 21, 1634, in Everett Emerson, ed., *Letters from New England* (Amherst, MA, 1976), pp. 119, 116.

15. Wood, *Prospect*, pp. 38-9. Edge habitats were of course disrupted not only by forest regrowth but by English agricultural clearing; either could have the same effect of reducing animal populations. (On the other hand, English fields sometimes acted as edge habitats themselves.)

16. Calvin Martin, "The European Impact on the Culture of a Northeastern Algonquian Tribe: An Ecological Interpretation," *William and Mary Quarterly*, 3rd ser., 31 (1974), pp. 3-26; Martin, *Keepers of the Game* (Berkeley, CA, 1978). Martin's work is discussed in Shepard Krech III, ed., *Indians, Animals, and the Fur Trade* (Athens, GA, 1981). One of Martin's more interesting arguments is that Europeans brought diseases not only to the Indians but to New World mammals generally; pathogens like tularemia could have created epizootics in which humans and animals literally did infect one another. I have found no evidence for this in New England, but would not rule it out as a possibility. See *Keepers*, pp. 130-49.

17. Wood, *Prospect*, p. 107. Europeans were not even much good at catching turkeys. Isaack de Rasieres said that "we generally take savages with us when we go to hunt them; for even when one has deprived them of the power of flying, they yet run so fast that we cannot catch them unless their legs are hit also." Isaack de Rasieres to Samuel Blommaert, ca. 1628, in Sydney V. James, Jr., ed., *Three Visitors to Early Plymouth* (Plimoth Plantation, 1963), pp. 79-80.

18. Williams, *Key*, p. 215; Morton, *Canaan*, p. 159; John Smith, *The Generall Historie of Virginia* (1624), reprinted Readex Microprint (1966), p. 215; Champlain, *Voyages*, p. 90. Bert Salwen (in "Indians of Southern

New England and Long Island: Early Period," in Bruce Trigger, ed., *Handbook of North American Indians, 15, Northeast,* 1978, p. 166) emphasizes the commercial nature of this exchange, which was certainly present; but it is equally important to see how such "commerce" diverged from European forms by being embedded in a different institutional framework.

19. Smith, *Generall Historie,* p. 215; Salwen, "Southern New England Indians," in Trigger, *Northeast,* p. 166; Thomas, *Maelstrom,* pp. 177-8; Jennings, *Invasion,* p. 103.

20. References to copper jewelry are Wroth, *Verrazzano,* p. 138; Brereton, "Relation," in Burrage, *Early Voyages,* p. 337; Martin Pring, "A Voyage Set Out from the Citie of Bristoll, 1602," in *ibid.,* p. 347; James Rosier, "A True Relation of the Voyage of Captaine George Waymouth, 1605," in *ibid.,* p. 373; Gabriel Archer, "Relation of Captain Gosnols Voyage to the North Part of Virginia, 1602," in Samuel Purchas, ed., *Purchas His Pilgrimes* (Glasgow, 1906), 18, pp. 305, 307. Champlain, *Voyages,* p. 90, gives an example of European goods entering into earlier Indian trade patterns. Thomas, *Maelstrom,* pp. 155-8, is good on aboriginal trade, and Susan G. Gibson, ed., *Burr's Hill* (Providence, 1980), pp. 72-6, 108-17, gives fine archaeological illustrations of trade goods that have been reworked by Indians.

21. The classic essay on gift exchange is Marcel Mauss, *The Gift* (New York, 1954). I have also used the discussions in Marshall Sahlins, *Stone Age Economics* (Chicago, 1972), pp. 149-314; Sahlins, *Tribesmen* (Englewood Cliffs, NJ, 1968), pp. 74-95; and George Dalton, "Aboriginal Economies in Stateless Societies," in Timothy K. Earle and Jonathan E. Ericson, eds., *Exchange Systems in Prehistory* (New York, 1977), pp. 191-212. American examples can be found in Wilbur R. Jacobs, *Diplomacy and Indian Gifts* (Stanford, 1950); and, for New England, Jennings, *Invasion.*

22. Bradford, *Plymouth,* p. 178; Champlain, *Voyages,* p. 90; Thomas, *Maelstrom,* pp. 172-5; James, *Early Visitors,* pp. 77-8; Roberts, *New England Fur Trade,* p. 33; Francis Higginson, *New-Englands Plantation* (1630), in *Massachusetts Historical Society Proceedings,* 62 (1929), p. 309.

23. Bradford, *Plymouth,* p. 203; Williams, *Key,* pp. 210-14; Frank G. Speck, "The Functions of Wampum Among the Eastern Algonkian," *Memoirs of the American Anthropological Association,* 6 (1919), pp. 56-71; Daniel Gookin, "Historical Collections of the Indians in New England," *Massachusetts Historical Society Collections,* 1st ser., 1 (1792), p. 152.

24. James, *Early Visitors,* p. 74; Bradford, *Plymouth,* pp. 203-4 (but see also p. 195, which demonstrates that the Pilgrims knew of wampum before de Rasieres' arrival, apparently without understanding its significance). Neil Salisbury, *Manitou and Providence* (New York, 1982), pp. 132-7, is good on wampum's new role as money. The Maine Indians' reluctance to accept wampum for two years (Bradford, p. 203) is interesting, since it probably indicates that circulation of

wampum was originally sanctioned only within certain trade spheres in New England, its movement controlled both by social rank and by geographical location. Breaching such spheres and opening up regional trade was one of the Europeans' most important effects.

25. Bradford, *Plymouth,* p. 203.

26. Williams, *Key,* p. 218; William Bradford, "A Descriptive and Historical Account of New England in Verse," *Massachusetts Historical Society Collections,* 1st ser., 3 (1794), p. 83. Note that Calvin Martin's arguments about the primary role of animals and disease in *pushing* Indians into trade cannot explain why trade *pulled* the wampum makers to produce more wampum.

27. Even the way a "fathom" of wampum was defined suggests the degree to which traditional Indian ways of valuing it had begun to interact with European monetary systems: it was the number of beads necessary to equal sixty English pence, and so fluctuated depending on exchange rates as colonial governments and London markets set them. William B. Weeden, "Indian Money as a Factor in New England Civilization," *Johns Hopkins University Studies in Historical and Political Science,* 2nd ser., 8-9 (1884), p. 18.

28. Nicolas Denys, *The Description and Natural History of the Coasts of North America (Acadia)* (1672), William F. Ganong, ed. (Toronto, Champlain Society Publications, II, 1908), p. 426; Champlain, *Voyages,* p. 90.

29. "Extract from an Indian History," *Massachusetts Historical Society Collections,* 1st ser., 9 (1804), p. 101; compare *The First Annual Report of the American Society for Promoting the Civilization and General Improvement of the Indian Tribes of the United States* (New Haven, 1824), p. 42.

30. Francis X. Moloney, *The Fur Trade in New England, 1620-1676* (Cambridge, MA, 1931), pp. 29, 44, 49-51, 67-77; Bradford, *Plymouth,* p. 319, note 4; Douglas Edward Leach, *The Northern Colonial Frontier, 1607-1713* (New York, 1966), pp. 92-4; Bernard Bailyn, *The New England Merchants in the Seventeenth Century* (Cambridge, MA, 1955), reprinted (New York, 1964), pp. 53-4.

31. Wood, *Prospect,* pp. 51-2; Thomas, *Maelstrom,* pp. 172-5; Nathaniel B. Shurtleff, ed., *Records of the Governor and Company of Massachusetts Bay in New England* (Boston, 1853), I, pp. 94, 236; Bradford, "New England in Verse," p. 82. On Indian adoption of European guns, see Patrick M. Malone, "Changing Military Technology among the Indians of Southern New England, 1600-1677," *American Quarterly,* 25 (1973), pp. 48-63.

32. John Josselyn, *New-Englands Rarities Discovered* (1672), Edward Tuckerman, ed., *Transactions and Collections of the American Antiquarian Society,* 4 (1860), p. 144; Wood, *Prospect,* p. 50; Glover M. Allen, "The Wild Turkey in New England," *Bulletin of the Essex County Ornithological Club,* 3:1 (December 1921), pp. 5-17; Samuel Deane, *The New England Farmer* (Worcester, MA, 1790), p. 291; Tim-

othy Dwight, *Travels in New England and New York* (1821), Barbara
Miller Solomon, ed. (Cambridge, MA, 1969), I, p. 35; Jeremy Belknap, *The History of New Hampshire* (Dover, NH, 1812), III, p. 125; A.
W. Schorger, *The Wild Turkey: Its History and Domestication* (Norman, OK, 1966), pp. 3-18. John Josselyn's assertion about passenger
pigeons in 1675 that "of late they are much diminished, the *English*
taking them with Nets" *(An Account of Two Voyages to New England*
[1675], in *Massachusetts Historical Society Collections*, 3rd ser., 3 [1833], p.
278) is a good example of a colonial source that must not be taken
at face value. The flights of pigeons actually cycled on an eleven-
or twelve-year basis, a fact of which Josselyn was unaware; he had
merely hit one of the low points in their cycle on his second journey. See A. W. Schorger, *The Passenger Pigeon* (Madison, 1955), pp.
205-8. Peter Matthiessen, *Wildlife in America* (New York, 1959, reprinted Baltimore, 1977), is good on later bird extinctions; see also
John and Mildred Teal, *Life and Death of the Salt Marsh* (New York,
1969).
33. Matthiessen, *Wildlife in America*, pp. 64-5; Dwight, *Travels*, I, p. 33;
George Waldo Browne, ed., *Early Records of Londonderry, Windham,
and Derry, N.H., 1719-1762* (Manchester, NH, 1908), pp. 242, 247, 249,
255, 267, 273, 278, 283, 291, 297, 300, 331, 334, 344, 374. Regulations of this
kind were of course being used at the same time in England to
restrict access to hunting to landowning classes, so that class conflict
as well as ecological changes may have contributed to their creation.
34. Gookin, "Historical Collections," p. 162. Note Gookin's reference to
wage labor, which was another important innovation among New
England Indians. Compare Shurtleff, *Massachusetts Records*, I, p. 83,
II, p. 152. On settlement shifts, see Lorraine E. Williams, *Ft. Shantok
and Ft. Corchaug: A Comparative Study of Seventeenth Century Culture
Contact in the Long Island Sound Area*, Ph.D. Thesis, New York Uni-
versity, 1972, p. 225; Williams, *Key*, p. 215; Lynn Ceci, *The Effect of
European Contact and Trade on the Settlement Pattern of Indians in
Coastal New York, 1524-1665: The Archeological and Documentary Evidence*,
Ph.D. Thesis, City University of New York, 1977, pp. 277ff; Bert
Salwen, "European Trade Goods and the Chronology of the Ft.
Shantok Site," *Bulletin of the Archaeological Society of Connecticut*, 34
(1966), pp. 36-7; Peter A. Thomas, "Squakheag Ethnohistory: A Pre-
liminary Study of Culture Conflict on the Seventeenth Century
Frontier," *Man in the Northeast*, 5 (1973), pp. 27-36.
35. Thomas, *Maelstrom*, pp. 183-8; Roberts, *New England Fur Trade*, p. 61;
Williams, *Key*, p. 187; Gookin, "Historical Collections," p. 152.
36. Gookin, "Historical Collections," p. 152; Weeden, "Indian Money";
Alden T. Vaughan, *New England Frontier: Puritans and Indians, 1620-
1675*, rev. ed. (New York, 1979), pp. 222-4; John R. Bartlett, ed., *Records
of the Colony of Rhode Island and Providence Plantation* (Providence,
1856), I, p 474. An interesting example of wampum's continued
circulation in the context of a very ad hoc monetary system on the

colonial frontier is in [Sarah Kemble Knight,] *The Journal of Madam Knight* (1704) (New York, 1935), pp. 40-1.

37. L. Williams, *Ft. Shantok*, pp. 49-50, 59, 147-52, 203; Gookin, "Historical Collections," pp. 180-93. Jennings, *Invasion*, is the best available account of Indian land losses in the seventeenth century.

38. Denys, *Acadia*, pp. 440, 442-52; Biard, *Relation*, p. 69. Good discussions of northern subsistence changes are in Alfred Goldsworthy Bailey, *The Conflict of European and Eastern Algonkian Cultures, 1504-1700*, 2nd ed. (Toronto, 1969); Cornelius J. Jaenen, *Friend and Foe: Aspects of French-Amerindian Cultural Contact in the Sixteenth and Seventeenth Centuries* (Toronto, 1976); Bruce J. Bourque, "Aboriginal Settlement and Subsistence on the Maine Coast," *Man in the Northeast*, 6 (Fall 1973), pp. 3-20; and Snow, "Abenaki Fur Trade."

39. Denys, *Acadia*, pp. 187, 441, 443; Calvin Martin, "The Four Lives of a Micmac Copper Pot," *Ethnohistory*, 22 (1975), pp. 111-33; James Sullivan, "The History of the Penobscott Indians," *Massachusetts Historical Society Collections*, 1st ser., 9 (1804), p. 228; Dwight, *Travels*, I, p. 33.

40. Joseph Chadwick, "An Account of a Journey from Fort Pownal— Now Fort Point—Up the Penobscot River to Quebec, in 1764," *Bangor Historical Magazine*, 4 (1889), p. 143. The dispute over Algonquian family hunting territories is of very long standing. I have taken a moderate position, arguing that some precolonial parceling of hunting territories probably did occur on an informal basis, one that changed from year to year, but that the fur trade crystallized the territories and directed their production toward market exchange. Dean R. Snow, "Wabanaki 'Family Hunting Territories,' " *American Anthropologist*, 70 (1968), pp. 1143-51, is closest to my position, and is a good review of the controversy; for more material, see the articles discussed in the bibliographical essay. Rene R. Gadacz, "Montagnais Hunting Dynamics in Historicoecological Perspective," *Anthropologica*, 17 (1975), pp. 149-67, argues for aboriginal conservation of game animals (something which Eleanor Leacock's classic "The Montagnais 'Hunting Territory' and the Fur Trade," *Memoirs of the American Anthropological Association*, 56:2, No. 78 [1954], denies) based on the principle of least effort; I side with Gadacz, but argue that market-oriented conservation is a much different phenomenon than conservation derived either from least effort or from Liebig's Law. In this, I am surprisingly close to Harold Demsetz, "Toward a Theory of Property Rights," *American Economic Review, Papers and Proceedings*, 57 (1967), pp. 347-59, with whose approach I otherwise differ.

41. Chadwick, "Journey to Quebec," p. 143; Benjamin Trumbull, *A Complete History of Connecticut* (Hartford, CT, 1797), p. 26; Dwight, *Travels*, I, 125; Samuel Williams, *The Natural and Civil History of Vermont*, 2nd ed. (Burlington, 1809), I, p. 121. My discussion does no justice to the complexity of the northern fur trade, located as it was

on the margins of English, French, Iroquois, and Abenaki spheres of influence. The effects of warfare in such circumstances were profound and fed into the ecological changes which I here merely adumbrate. See Leach, *Northern Colonial Frontier,* and Kenneth M. Morrison, *The People of the Dawn: The Abnaki and Their Relations with New England and New France, 1600-1727,* Ph.D. Thesis, University of Maine, 1975, for more detailed examinations of political interactions in the area.

42. H. L. Babcock, "The Beaver as a Factor in the Development of New England," *Americana,* 11 (1916), p. 185; compare Winthrop, *Journal,* I, p. 73; and Henry Wansy, *An Excursion to the United States of America in the Summer of 1794,* 2nd ed. (Salisbury, 1798), p. 197. My discussion of the ecological consequences of beaver dam removal is derived from Belknap, *History of New Hampshire,* III, pp. 58-9, 113-19. See also Jared Eliot, *Essays upon Field Husbandry in New England and Other Papers, 1748-1762,* Harry J. Carman and Rexford G. Tugwell, eds. (New York, 1934), p. 14; George Perkins Marsh, *Man and Nature* (1864), David Lowenthal, ed. (Cambridge, MA, 1965), p. 32; Austin F. Hawes, "New England Forests in Retrospect," *Journal of Forestry,* 21 (1923), p. 215; and S. A. Wilde, *et al.,* "Changes in Composition of Ground Water, Soil Fertility, and Forest Growth Produced by the Construction and Removal of Beaver Dams," *Journal of Wildlife Management,* 14 (1950), pp. 123-8.

43. Dwight, *Travels,* I, p. 33; "Harry Quaduaquid and Robert Ashpo to the Most Honourable Assembly of the State of Connecticut" (May 14, 1789), quoted in Matthiessen, *Wildlife in America,* pp. 70-1.

6. Taking the Forest

1. William Bradford, *Of Plymouth Plantation,* Samuel Eliot Morison, ed. (New York, 1952), p. 94.

2. William Douglass, *A Summary, Historical and Political, of the First Planting, Progressive Improvements, and Present State of the British Settlements in North America* (London, 1760), II, pp. 52-72; Robert G. Albion, *Forests and Sea Power: The Timber Problem of the Royal Navy, 1652-1862* (Cambridge, MA, 1926), pp. 3-38.

3. Samuel Maverick, "A Briefe Discription of New England," *Massachusetts Historical Society Proceedings,* 2nd ser., 1 (1884-5), p. [233]; Austin F. Hawes, "New England Forests in Retrospect," *Journal of Forestry,* 21 (1923), p. 215; Samuel Williams, *The Natural and Civil History of Vermont,* 2nd ed. (Burlington, 1809), I, p. 87; George Perkins Marsh, *Man and Nature* (1864), David Lowenthal, ed. (Cambridge, MA, 1965), pp. 236-7; Charles F. Carroll, *The Timber Economy of Puritan New England* (Providence, 1973), pp. 35-7, 86; William R. Carlton, "New England Masts and the King's Navy," *New England Quarterly,* 12

(1939), pp. 4-6; Joseph J. Malone, *Pine Trees and Politics: The Naval Stores and Forest Policy in Colonial New England, 1691-1775* (Seattle, WA, 1964), p. 3; John Winthrop, *Winthrop's Journal,* James Kendall Hosmer, ed. (New York, 1908), I, p. 129; Richard M. Candee, "Merchant and Millwright: The Water Powered Sawmills of the Piscataqua," *Old-Time New England,* 60 (1970), pp. 131-49; James Elliott Defebaugh, *History of the Lumber Industry of America* (Chicago, 1907), II, pp. 15, 21; John E. Hobbs, "The Beginnings of Lumbering as an Industry in the New World, and First Efforts at Forest Protection," *Forest Quarterly,* 4 (1906), pp. 14-23.

4. J. P. Kinney, "Forest Legislation in America Prior to March 4, 1789," *Cornell University Agricultural Experiment Station Bulletin,* 370 (1916), pp. 389-96; Carroll, *Timber Economy,* pp. 116-19; Carlton, "New England Masts," pp. 9-10; Albion, *Forests and Sea Power,* pp. 231-80; Malone, *Pine Trees.*

5. J. Willcox Brown, "Forest History of Mount Moosilauke," *Appalachia,* 24 (1958), p. 26; Defebaugh, *Lumber Industry,* II, p. 138.

6. Albion, *Forests and Sea Power,* pp. 17-26; Paul Huffington and J. Nelson Clifford, "Evolution of Shipbuilding in Southeastern Massachusetts," *Economic Geography,* 15 (1939), pp. 363-9; Richard S. Dunn, *Sugar and Slaves: The Rise of the Planter Class in the English West Indies, 1624-1713* (Chapel Hill, 1972), pp. 29-34, 67; Carroll, *Timber Economy,* pp. 77-84; J. Hammond Trumbull, ed., *The Public Records of the Colony of Connecticut* (Hartford, 1850), I, p. 60; Peter Kalm, *Travels in North America* (1753-61, 1770), Adolph B. Benson, ed. (New York, 1964), pp. 20, 298-301; Anthony N. B. Garvan, *Architecture and Town Planning in Colonial Connecticut* (New Haven, 1951), p. 99; Daniel Gookin, "Historical Collections of the Indians in New England," *Massachusetts Historical Society Collections,* 1st ser., 1 (1792), p. 184.

7. Timothy Dwight, *Travels in New England and New York* (1821), Barbara Miller Solomon, ed. (Cambridge, MA, 1969), IV, p. 151.

8. *Ibid.,* II, pp. 72, 138-9; James Birket, *Some Cursory Remarks* (1750-1) (New Haven, 1916), pp. 4, 12; Jeremy Belknap, *The History of New Hampshire* (Dover, NH, 1812), III, p. 150; Peter Whitney, *The History of the County of Worcester* (Worcester, MA, 1793), pp. 88, 124-5, 146, 173, 187, 190, 214, 267, 269, 297, 302.

9. Kalm, *Travels,* I, p. 300; Whitney, *Worcester,* p. 191; "On the Importance of Preserving Forests in the United States," *Weekly Magazine,* 2:16 (May 19, 1798), p. 80; Justin Winsor, *A History of the Town of Duxbury, Massachusetts* (Boston, 1849), p. 34; Dwight, *Travels,* I, pp. 74-5; J. Gordon Ogden, III, "Forest History of Martha's Vineyard, Massachusetts: I. Modern and Pre-Colonial Forests," *American Midland Naturalist,* 66 (1961), p. 428.

10. "On the Importance of Preserving Forests"; B[enjamin] Lincoln, "Remarks on the Cultivation of the Oak," *Massachusetts Historical Society Collections,* 2nd ser., 1 (1814), pp. 188-9; Philip T. Coolidge, *History of the Maine Woods* (Bangor, ME, 1963), pp. 31-2.

11. Dwight, *Travels*, III, pp. 176-7; Jedidiah Morse, *The American Geography*, 2nd ed. (London, 1792), p. 143; Douglass, *Present State of Settlements in North America*, II, p. 216; Belknap, *History of New Hampshire*, III, pp. 95-7.

12. Dwight, *Travels*, IV, p. 72.

13. Belknap, *History of New Hampshire*, III, p. 97. Descriptions of this technique are fairly frequent in the colonial literature, especially in the writings of European travelers who were appalled at the messy fields it created.

14. Dwight, *Travels*, II, pp. 83-4; Jared Eliot, *Essays upon Field Husbandry in New England* (1748), Harry J. Carman and Rexford G. Tugwell, eds. (New York, 1934), p. 7.

15. Samuel Deane, *The New England Farmer* (Worcester, MA, 1790), p. 57; Dwight, *Travels*, II, pp. 325-6.

16. R. Dossie, *Observations on the Pot-Ash Brought from America* (London, 1767); Hawes, "New England Forests," pp. 217-19; Robert Montgomery, *Discourse Concerning the Establishment of a New Colony to the South of Carolina* (1717), in Peter Force, ed., *Tracts* (Washington, 1838), I, pp. 20-21; Harry Miller, "Potash from Wood Ashes," *Technology and Culture*, 21 (1980), pp. 187-208.

17. John Winthrop, *Journal*, James K. Hosmer, ed. (New York, 1908), I, p. 54; Kinney, "Forest Legislation," pp. 363-7.

18. Kinney, "Forest Legislation," p. 366.

19. John Robert Stilgoe, *Pattern on the Land: The Making of a Colonial Landscape, 1633-1800*, Ph.D. Thesis, Harvard University, 1976, pp. 98-129; Benno M. Forman, "Mill Sawing in Seventeenth-Century Massachusetts," *Old-Time New England*, 60 (1970), pp. 110-30.

20. Garvan, *Connecticut Architecture*, pp. 78-102; Carl Bridenbaugh, "Yankee Use and Abuse of the Forest in the Building of New England, 1620-1660," *Massachusetts Historical Society Proceedings*, 89 (1977), pp. 3-35.

21. Jacques Pierre Brissot de Warville, *On America* (1792), reprinted (New York, 1970), I, pp. 127-8; Howard S. Russell, *A Long, Deep Furrow: Three Centuries of Farming in New England* (Hanover, NH, 1976), pp. 185-6; Hector St. John de Crèvecœur, *Sketches of Eighteenth-Century America*, Henri L. Bourdin, *et al.*, eds. (New Haven, 1925), p. 81; Kalm, *Travels*, I, pp. 50, 77, 238-9; Dwight, *Travels*, IV, p. 152; Lewis M. Norton, *Goshen in 1812* (Acorn Club of Connecticut, 1949), p. 12; Percy Wells Bidwell, "Rural Economy in New England at the Beginning of the Nineteenth Century," *Transactions of the Connecticut Academy of Arts and Sciences*, 20 (1916), p. 335, note 4.

22. Francis Higginson, *New-Englands Plantation* (1630), *Massachusetts Historical Society Proceedings*, 62 (1929), p. 314; Crèvecœur, *Sketches*, p. 72; Benjamin Rush, *Essays, Literary, Moral and Philosophical*, 2nd ed. (Philadelphia, 1806), p. 229; Kalm, *Travels*, I, p. 239; Hawes, "New England Forests," p. 220; Stilgoe, *Pattern on the Land*, p. 61; R. V. Reynolds and Albert H. Pierson, "Fuel Wood Used in the United States, 1630-1930," *USDA Circular*, 641 (February 1942), pp. 9, 14.

23. Nathaniel B. Shurtleff, ed., *Records of the Governor and Company of the Massachusetts Bay in New England* (Boston, 1853), I, pp. 104, 129; Winthrop, *Journal*, I, p. 258; Kinney, "Forest Legislation," pp. 371-7; Joseph Hadfield, *An Englishman in America* (1785), Douglas S. Robertson, ed. (Toronto, 1933), p. 216; Dwight, *Travels*, I, p. 125; II, p. 238; IV, pp. 150-1; Alfred P. Muntz, *Changing Geography of the New Jersey Woodland*, Ph.D. Thesis, University of Wisconsin, 1964; Douglas R. McManis, *Colonial New England* (New York, 1975), pp. 113-16; William Strickland, *Journal of a Tour in the United States of America, 1794-1795*, N. E. Strickland, ed. (New York, 1971), pp. 203, 212; Strickland, *Observations on the Agriculture of the United States of America* (1801), reprinted in *ibid.*, p. 6.

24. Kalm, *Travels*, I, p. 308.

25. Colonial writers tended to exaggerate the effects of tree removal on weather, asserting with Edward Johnson that clearing resulted in "changing the very nature of the seasons, moderating the Winters cold of late very much." (Edward Johnson, *Johnson's Wonder-Working Providence* [1654], J. Franklin Jameson, ed. [New York, 1910], p. 84.) For the most intelligent colonial discussion arguing that deforestation did in fact alter wind patterns, see Dwight, *Travels*, I, pp. 38-48; the Dwight quotation I use is on p. 41. My own discussion of the climatic and hydrological effects of deforestation relies on two modern studies to confirm what I take from the colonial sources: Joseph Kittredge, *Forest Influences: The Effects of Woody Vegetation on Climate, Water, and Soil* (New York, 1948); and Richard Lee, *Forest Hydrology* (New York, 1980).

26. Williams, *History of Vermont*, pp. 70-76; Kittredge, *Forest Influences*, pp. 52-71, 146-214; Lee, *Forest Hydrology*, pp. 154-81.

27. Williams, *History of Vermont*, p. 70; Dwight, *Travels*, I, p. 40.

28. Herbert I. Winer, *History of the Great Mountain Forest, Litchfield County, Connecticut*, Ph.D. Thesis, Yale University, 1955, p. 26; Lee, *Forest Hydrology*, pp. 182-216, 281-6; Belknap, *History of New Hampshire*, pp. 52-3; Dwight, *Travels*, II, p. 60; Sidney Perley, *Historic Storms of New England* (Salem, MA, 1891), pp. 1-167.

29. Noah Webster, *A Collection of Essays and Fugitive Writings on Moral, Historical, Political and Literary Subjects* (Boston, 1790), pp. 371-2.

30. Kittredge, *Forest Influences*, pp. 215-29; Lee, *Forest Hydrology*, pp. 111-29, 154-81; Dwight, *Travels*, IV, pp. 77-8; Webster, *Essays*, p. 371; Carroll, *Timber Economy*, p. 61; John Duffy, *Epidemics in Colonial America* (Baton Rouge, LA, 1953), pp. 202-14.

31. Ira Allen, *The Natural and Political History of the State of Vermont* (London, 1798), pp. 11-12; Marsh, *Man and Nature*, p. 172; Whitney, *Worcester*, p. 179; Defebaugh, *Lumber Industry*, II, p. 16; Williams, *History of Vermont*, p. 75.

32. A fine summary of these effects is in Marsh, *Man and Nature*, pp. 186-7.

7. A World of Fields and Fences

1. Manasseh Minor, *The Diary of Manasseh Minor* (1915); Thomas Tusser, *Five Hundred Points of Good Husbandrie* (London, 1580); *The Husbandman's Guide*, 2nd ed. (New York, 1712), pp. 3-15; Edwin Stanley Welles, *The Beginnings of Fruit Culture in Connecticut* (Hartford, CT, 1936), pp. 30-2; Darrett B. Rutman, *Husbandmen of Plymouth* (Boston, 1967), pp. 50-2.
2. Conrad M. Arensberg, "The Old World Peoples," *Anthropological Quarterly*, 36 (1963), pp. 75-99.
3. Compare John Winthrop, *Winthrop's Journal*, James Kendall Hosmer, ed. (New York, 1908); and William Bradford, *Of Plymouth Plantation*, Samuel Eliot Morison, ed. (New York, 1952). (The quotation is on p. 141.) Other quotations are in Everett Emerson, ed., *Letters from New England* (Amherst, MA, 1976), pp. 110, 225; and William Wood, *New England's Prospect* (1634), Alden T. Vaughan, ed. (Amherst, MA, 1977), p. 69. Plymouth had other livestock before it first obtained cattle: in 1623, it possessed six goats, fifty swine, and a number of hens. Emmanuel Altham to Sir Edward Altham, September 1623, in *Three Visitors to Early Plymouth*, Sidney V. James, Jr., ed. (Plimoth Plantation, 1963), p. 24.
4. Carl Bridenbaugh, *Fat Mutton and Liberty of Conscience* (Providence, 1974), pp. 27-60; Percy Wells Bidwell and John I. Falconer, *History of Agriculture in the Northern United States* (Washington, D.C., 1925), pp. 18-32; Howard S. Russell, *A Long, Deep Furrow: Three Centuries of Farming in New England* (Hanover, NH, 1976), pp. 30-8, 151-69.
5. Winthrop, *Journal*, I, p. 64; J. Hammond Trumbull, ed., *The Public Records of the Colony of Connecticut* (Hartford, 1850), p. 19; Nathaniel B. Shurtleff, ed., *Records of the Governor and Company of the Massachusetts Bay in New England* (Boston, 1853), IV, Part 2, pp. 512-13.
6. John Winthrop, "Reasons to Be Considered," *Winthrop Papers*, Massachusetts Historical Society (1931), II, p. 141.
7. Shurtleff, *Massachusetts Records*, I, pp. 121, 133; Franklin Bowditch Dexter, ed., *New Haven Town Records, I, 1649-1662* (New Haven, CT, 1917), p. 193; Nathaniel B. Shurtleff, ed., *Records of the Colony of New Plymouth in New England* (Boston, 1855), III, pp. 21, 89, 106, 119, 132, 192; James P. Ronda, "Red and White at the Bench: Indians and the Law in Plymouth Colony, 1620-1691," *Essex Institute Historical Collections*, 110 (1974), pp. 208-9.
8. Shurtleff, *Plymouth Records*, III, p. 192; Dexter, *New Haven Records*, I, p. 193; Shurtleff, *Massachusetts Records*, I, p. 99.
9. I am indebted to Edmund Morgan for the suggestion that colonial wolf populations probably rose after the arrival of English livestock; the anonymous author of the "Essay on the Ordering of Towns" (in the *Winthrop Papers*, III, p. 185) corroborates this, but for different reasons, with the claim that "I have often hearde (by seemeing credible men) that Wolves are much more increased since

our Nation came then when the Indians possessed the same, and a Reason rendred, that they were dilligent in destroying the Yonge." One can doubt the "Reason rendred" by wondering whether a people who kept no livestock would have troubled themselves so much over predators who lived off the deer herds, but perhaps they did. Colonial wolf populations are impossible to estimate accurately. One gets the feeling that, for colonists, wolves were either "very common, and very noxious," or were nonexistent. There does not appear to have been any middle ground between these two conditions. Colonists, like many who keep cattle today, surely overestimated the damage done by wolves, and probably attributed to wolves livestock deaths which had nothing to do with the predators. On early responses to wolves, see Shurtleff, *Massachusetts Records*, III, pp. 10, 17; Winthrop, *Journal*, I, pp. 53, 67, III; Wood, *Prospect*, pp. 45-6.

10. On wolf bounties, see Shurtleff, *Massachusetts Records*, I, pp. 81, 102, 156, 218, 252, 304, 319; II, pp. 84-5, 103, 252; III, pp. 10, 17, 134, 319; IV, Part 2, pp. 2, 42; V, p. 453; Shurtleff, *Plymouth Records*, I, pp. 22, 31; III, pp. 50-1, 85-6; *Town Records of Salem* (Salem, 1868), pp. 107, 127. Most town records contain a number of entries similar to the ones I cite here.

11. John Josselyn, *New-England's Rarities Discovered* (1672), Edward Tuckerman, ed., in *Transactions and Collections of the American Antiquarian Society*, 4 (1860), pp. 150-1; Dexter, *New Haven Records*, pp. 73-4, 92, 309; Shurtleff, *Massachusetts Records*, II, pp. 252-3; M. Minor, *Diary*, pp. 32, 48, 53, 81, 98, 105, 113, 119, 120; Jeremy Belknap, *History of New Hampshire* (Dover, NH, 1812), III, pp. 108-9; Benjamin Trumbull, *A Complete History of Connecticut* (Hartford, CT, 1797), p. 26; Herbert B. Adams, "Village Communities of Cape Anne and Salem," *Johns Hopkins University Studies in Historical and Political Science*, ser. 1, 9-10 (1883), p. 58; "Ordering of Towns," *Winthrop Papers*, III, p. 184; *Scituate Records*, I, p. 48, as quoted by John Robert Stilgoe, *Pattern on the Land: The Making of a Colonial Landscape, 1633-1800*, Ph.D. Thesis, Harvard University, 1976, p. 159; Timothy Dwight, *Travels in New England and New York* (1821), Barbara Miller Solomon, ed. (Cambridge, MA, 1969), I, p. 33.

12. Shurtleff, *Plymouth Records*, I, p. 6; Shurtleff, *Massachusetts Records*, I, pp. 215, 221, 272; II, pp. 14-15.

13. Charles J. Hoadly, ed., *Records of the Colony of New Haven* (Hartford, 1858), p. 579; Dexter, *New Haven Records*, pp. 65, 132, 234, 281; Shurtleff, *Massachusetts Records*, I, p. 333; III, p. 319; David Thomas Konig, "Community Custom and the Common Law: Social Change and the Development of Land Law in Seventeenth-Century Massachusetts," *American Journal of Legal History*, 18 (1974), pp. 137-77. For a fine detailed discussion of how earlier English field practices fed into these colonial systems, see David Grayson Allen, *In English Ways* (Chapel Hill, 1981).

14. Rutman, *Husbandmen of Plymouth*, p. 49; Shurtleff, *Massachusetts Rec-*

ords, I, pp. 106, 150, 157, 182, 219-20, 222, 238-9, 255, 265, 270; IV, Part 2, p. 322. It is quite likely that disputes over swine expressed a disguised class conflict. Because pigs were so cheap and easy to raise, they were favored by poorer colonists as a source of meat; wealthier colonists, who could afford to keep larger numbers of cattle, had less need of them. The evidence cited above in Shurtleff suggests that a number of colonists were decidedly unhappy about the swine laws, and spoke against them so vociferously that the Massachusetts Court felt compelled to mete out stiff fines. No study of this issue has been done for colonial New England, but Steven Hahn's article on the nineteenth-century South is suggestive: "Hunting, Fishing, and Foraging: Common Rights and Class Relations in the Postbellum South," *Radical History Review*, 26 (October 1982), pp. 37-64.

15. Shurtleff, *Massachusetts Records*, I, pp. 188-9; Roger Williams, *The Letters of Roger Williams*, John Russell Bartlett, ed., *Publications of the Narragansett Club*, 6 (1874), pp. 71, 78, 104; "Leift Lion Gardener His Relation of the Pequot Warres," *Massachusetts Historical Society Collections*, 3rd ser., 3 (1833), p. 154; Roger Williams, *A Key into the Language of America* (1643), John J. Teunissen and Evelyn J. Hinz, eds. (Detroit, 1973), p. 182; Wood, *Prospect*, p. 57; Thomas Morton, *New English Canaan* (1637), Charles F. Adams, ed., *Pubs. of the Prince Society*, XIV (Boston, 1883), p. 227.

16. *Salem Records*, pp. 130, 137, 143, 152; Dexter, *New Haven Records*, pp. 19-20; Shurtleff, *Massachusetts Records*, I, p. 215.

17. On these land divisions, see the general list of town studies and histories of New England agriculture in the bibliographical essay.

18. On speculation, see the bibliographical essay.

19. George Perkins Marsh, *Man and Nature* (1864), David Lowenthal, ed. (Cambridge, MA, 1965), p. 74; Allen, *In English Ways*, p. 231; Hugh M. Raup and Reynold E. Carlson, "The History of Land Use in the Harvard Forest," *Harvard Forest Bulletin*, 20 (1941), p. 25.

20. Richard Bushman, *From Puritan to Yankee* (Cambridge, MA, 1967), pp. 31-2; Bridenbaugh, *Fat Mutton*, pp. 27-60; Bidwell and Falconer, *Northern Agriculture*, pp. 26-32, 40-8; Darrett B. Rutman, "Governor Winthrop's Garden Crop: The Significance of Agriculture in the Early Commerce of Massachusetts Bay," *William and Mary Quarterly*, 3rd ser., 20 (1963), pp. 396-415; Samuel Maverick, "A Briefe Discription of New England," *Massachusetts Historical Society Proceedings*, 2nd ser., 1 (1884-85), p. 247; Joseph Hadfield, *An Englishman in America, 1785*, Douglas S. Robertson, ed. (Toronto, 1933), p. 198.

21. M. Minor, *Diary*, pp. 31, 42, 49, 50, 56, 57, 74, etc.; Carl Bridenbaugh, "Yankee Use and Abuse of the Forest in the Building of New England, 1620-1666," *Massachusetts Historical Society Proceedings*, 89 (1977), pp. 34-5; *American Husbandry* (1775), Harry J. Carman, ed. (New York, 1939), pp. 44-5; Gary B. Nash, *The Urban Crucible* (Cambridge, MA, 1979).

22. E. Fraser Darling, "Man's Ecological Dominance through Domes-

ticated Animals on Wild Lands," in William L. Thomas, ed., *Man's Role in Changing the Face of the Earth* (Chicago, 1956), p. 781.

23. Winthrop, *Journal*, I, pp. 132, 151; Emerson, *Letters*, p. 154; Bradford, *Plymouth Plantation*, p. 253.

24. Thomas Hutchinson, *The History of the Colony and Province of Massachusetts-Bay* (1765), Lawrence Shaw Mayo, ed. (Cambridge, MA, 1936), I, p. 405; John Smith, "Advertisements for the Unexperienced Planters of New-England" (1631), *Massachusetts Historical Society Collections*, 3rd ser., 3 (1833), p. 37; Emerson, *Letters*, pp. 214, 227.

25. Wood, *Prospect*, p. 34; James, *Plymouth Visitors*, p. 67; Everett Edwards, "The Settlement of Grasslands," in *USDA Yearbook, Grass* (1948), p. 17; Bidwell and Falconer, *Northern Agriculture*, p. 20; Lyman Carrier, *The Beginnings of Agriculture in America* (New York, 1923), pp. 239-43; Bridenbaugh, *Fat Mutton*, pp. 31-3; Roger Williams, *Letters*, John R. Bartlett, ed. (Providence, 1874), pp. 146-7; Robert R. Walcott, "Husbandry in Colonial New England," *New England Quarterly*, 9 (1936), pp. 239-40. At least one New England town—New Haven— made an effort to protect its English grasses during the early years of settlement. The town voted in 1654: "All men were desired to take notice that if any cut up any English grass which growes about the markit place, the streets, or other commons, to plant in their owne ground, they must expect to receive due punishment for the same." Dexter, *New Haven Records*, p. 204.

26. Alfred J. Crosby, "Ecological Imperialism: The Overseas Migration of Western Europeans as a Biological Phenomenon," *Texas Quarterly*, 30 (1978), pp. 18-19; Hutchinson, *History of Massachusetts-Bay*, I, p. 403; Herbert G. Baker, "The Evolution of Weeds," *Annual Review of Ecology and Systematics*, 5 (1974), p. 4; Margaret B. Davis, "Phytogeography and Palynology of Northeastern United States," in H. E. Wright, Jr., and David G. Grey, eds., *The Quaternary of the United States* (Princeton, NJ, 1965), p. 396; Richard B. Brugam, "Pollen Indicators of Land-Use Change in Southern Connecticut," *Quaternary Research*, 9 (1978), pp. 349-62.

27. John Josselyn, *New-England's Rarities Discovered* (1672), Edward Tuckerman, ed., *Transactions and Collections of the American Antiquarian Society*, 4 (1860), pp. 216-19; Asa Gray, "The Flora of Boston and Its Vicinity, and the Changes It Has Undergone," in Justin Winsor, ed., *The Memorial History of Boston* (Boston, 1880), pp. 17-22; Gray, "The Pertinacity and Predominance of Weeds," *American Journal of Science and Arts*, 3rd ser., 18:105 (September 1879), pp. 161-7; Dexter, *New Haven Records*, p. 132. I have relied throughout on Merritt L. Fernald, *Gray's Manual of Botany*, 8th ed. (New York, 1950), to determine whether a plant is of European or American origins. An interesting popular account of plant migrations is Claire S. Houghton, *Green Immigrants: The Plants That Transformed America* (New York, 1978).

28. Jared Eliot, *Essays upon Field Husbandry in New England* (1748-62),

Harry J. Carman and Rexford G. Tugwell, eds. (New York, 1934), pp. 27-9, 61-6; Carrier, *Beginnings of Agriculture*, pp. 239-42; Bidwell and Falconer, *Northern Agriculture*, pp. 103-5; Samuel Deane, *The New-England Farmer* (Worcester, MA, 1790), pp. 28-9, 285-6.

29. Peter Whitney, *History of the County of Worcester* (Worcester, MA, 1793), p. 203; Harold J. Lutz, "Trends and Silvicultural Significance of Upland Forest Successions in Southern New England," *Yale School of Forestry Bulletin*, 22 (1928), p. 22; Stanley W. Bromley, "The Original Forest Types of Southern New England," *Ecological Monographs*, 5 (1935), pp. 79-80; Eliot, *Essays*, p. 19.

30. Lutz, "Trends of Upland Forest Succession," p. 22; H. I. Winer, *History of the Great Mountain Forest, Litchfield County, Connecticut*, Ph.D. Thesis, Yale University, 1955, p. 255; Bromley, "Original Forest Types," p. 80; Timothy Dwight, *Travels in New England and New York* (1821), Barbara M. Solomon, ed. (Cambridge, MA, 1969), II, pp. 309-10; P. L. Marks, "The Role of Pin Cherry *(Prunus pensylvanica* L.) in the Maintenance of Stability in Northern Hardwood Ecosystems," *Ecological Monographs*, 44 (1974), pp. 73-88.

31. Dwight, *Travels*, I, p. 75; Bromley, "Original Forest Types," pp. 75, 80; E. Lucy Braun, *Deciduous Forests of Eastern North America* (New York, 1950), p. 253; G. E. Nichols, "The Vegetation of Connecticut, II, Virgin Forests," *Torreya*, 13 (1913), pp. 199-215; Lutz, "Trends of Upland Forest Succession," p. 15.

32. B[enjamin] Lincoln, "Remarks on the Cultivation of the Oak," *Massachusetts Historical Society Collections*, 2nd ser., 1 (1814), p. 193.

33. E. A. Johnson, "Effects of Farm Woodland Grazing on Watershed Values in the Southern Appalachian Mountains," *Journal of Forestry*, 50 (1952), pp. 109-13; Harry O. Buckman and Nyle C. Brady, *The Nature and Property of Soils*, 7th ed. (New York, 1969), pp. 249-53; Eugene P. Odum, *Fundamentals of Ecology*, 3rd ed. (Philadelphia, 1971), pp. 418-19.

34. Gottfried Pfeifer, "The Quality of Peasant Living in Central Europe," in Thomas, *Man's Role in Changing the Earth*, pp. 249-53.

35. Eliot, *Essays*, p. 204; Angus McDonald, *Early American Soil Conservationists* (1941), *USDA Miscellaneous Publications*, #449 (Washington, D.C., 1971), pp. 3-19; F. H. Bormann, *et al.*, "The Export of Nutrients and Recovery of Stable Conditions Following Deforestation at Hubbard Brook," *Ecological Monographs*, 44 (1974), pp. 255-77. The literature about Hubbard Brook, on which much of the argument of this paragraph relies, is quite large; see the bibliographical essay for further references.

36. Brugam, "Pollen Indicators," p. 357; Ronald B. Davis, *et al.*, "Vegetation and Associated Environments During the Past 14,000 Years Near Moulton Pond, Maine," *Quaternary Research*, 5 (1975), pp. 439-40; Deane, *New-England Farmer*, p. 153.

37. New Haven in 1641 [Brockett], New Haven, copied in pen and ink from the original (now lost), by Mrs. Sarah C. D. Woodward, 1880-81;

Map of New Haven, drawn by Joseph Brown, 1724, copy made 1881 by Mrs. S. C. D. Woodward; Plan of the City of New Haven, June 6, 1802, New Haven, 1802; all of the above are in the map collection of Sterling Library, Yale University. Dwight, *Travels*, I, p. 131; Walter Muir Whitehill, *Boston: A Topographical History* (Cambridge, MA, 1968), pp. 1-46; Jedidiah Morse, *The American Geography* (London, 1792), p. 169; Samuel de Champlain, *Voyages*, Charles Pomeroy Otis, ed. (Boston, 1878), II, p. 82n.

38. Dwight, *Travels*, II, pp. 363-4; III, pp. 60-1; Clarence E. Olmstead, "Vegetation of Certain Sand Plains of Connecticut," *Botanical Gazette*, 99 (1937), pp. 270-4; J. P. Kinney, "Forest Legislation in America Prior to March 4, 1789," *Cornell University Agricultural Experiment Station Bulletin*, 370 (1916), pp. 397-8; Justin Winsor, *A History of the Town of Duxbury, Massachusetts* (Boston, 1849), p. 28; "A Topographical Description of Truro," *Massachusetts Historical Society Collections*, 1st ser., 3 (1794), p. 198; John R. Stilgoe, "A New England Coastal Wilderness," *Geographical Review*, 71 (1981), p. 36; Ralph H. Brown, *Historical Geography of the United States* (New York, 1948), pp. 27-9.

39. Alfred Crosby, *The Columbian Exchange* (Westport, CT), 1972, p. 175; *American Husbandry*, p. 39; Hutchinson, *History of Massachusetts*, I, p. 404. On the issue of colonial perceptions of soil quality, compare John Pratt, "Pratt's Apology," *Massachusetts Historical Society Collections*, 2nd ser., 7 (1818), pp. 126-9.

40. [Sarah Kemble Knight], *The Journal of Madam Knight* (1704) (New York, 1935), p. 63; Deane, *New-England Farmer*, pp. 179-80; Bidwell and Falconer, *Northern Agriculture*, p. 85; *American Husbandry*, pp. 58-9.

41. Shurtleff, *Massachusetts Records*, I, p. 114; Dwight, *Travels*, II, pp. 359, 362; "Letters on Indian Corn Cultivation," *Science*, 189 (September 19, 1975), p. 946; G. Browne Goode, "The Use of Agricultural Fertilizers by the Amerian Indians and the Early English Colonists," *American Naturalist*, 14 (1880), p. 474.

42. Dwight, *Travels*, I, p. 164; III, p. 214; James Birket, *Some Cursory Remarks* (1750-51) (New Haven, 1916), p. 7; James Sullivan, *The History of the District of Maine* (Boston, 1795), p. 21; Trumbull, *History of Connecticut*, p. 18.

43. William Hubbard, *A General History of New England* (circa 1682), *Massachusetts Historical Society Collections*, 2nd ser., 5 (1815), pp. 22-3; Eliot, *Essays*, pp. 17-19, 43-7; Dwight, *Travels*, III, p. 59.

44. Deane, *New-England Farmer*, p. 151; Dwight, *Travels*, I, p. 31; II, p. 238; III, pp. 210-11; Bidwell and Falconer, *Northern Agriculture*, pp. 95-6.

45. Thomas, *Man's Role*, pp. 799-800; Crosby, *Columbian Exchange*, p. 97; Crosby, "Ecological Invasion," pp. 15-6; Belknap, *History of New Hampshire*, III, pp. 119, 136; John Josselyn, *An Account of Two Voyages to New-England* (1675), *Massachusetts Historical Society Collections*, 3rd

ser., 3 (1833), p. 292; Bidwell and Falconer, *Northern Agriculture*, p. 32;
Peter Matthiessen, *Wildlife in America* (New York, 1959), p. 64.
Dwight *(Travels,* I, p. 36) thought the honeybee to be a native.
46. H. E. Wright and M. L. Heinselman, "Introduction" (to special
issue on fire ecology), *Quaternary Research,* 3 (1973), p. 322; Williams,
Key, p. 191; Winthrop, *Journal,* II, pp. 91-2, 277; Walcott, "Colonial
Husbandry," pp. 233-5; Deane, *New-England Farmer,* pp. 147-52;
Dwight, *Travels,* I, pp. 28, 51-2; Peter Kalm, *Travels in North America*
(1753-61, 1770), Adolph B. Benson, ed. (New York, 1964), I, p. 166.
47. John Hull, "Memoir and Diaries," *Transactions and Collections of the
American Antiquarian Society,* 3 (1857), p. 213; Walcott, "Colonial Husbandry," p. 235; Bidwell and Falconer, *Northern Agriculture,* pp.
93-4; Deane, *New-England Farmer,* pp. 249-54; Hutchinson, *History
of Massachusetts,* I, p. 406; Dwight, *Travels,* I, pp. 53, 276-8; II, pp.
235-8; Andrew McFarland Davis, "Barberry Bushes and Wheat,"
Publications of the Colonial Society of Massachusetts, 11 (1907), pp. 73-94;
G. L. Carefoot and E. R. Sprott, *Famine on the Wind: Plant Diseases
and Human History* (London, 1969), pp. 38-41.
48. Eliot, *Essays,* pp. 7-52; Dwight, *Travels,* I, pp. 234-6, 246; II, pp. 2-3,
359-60; Marquis de Chastellux, *Travels in North America in the Years
1780, 1781 and 1782* (1786), Howard C. Rice, Jr., ed. (Chapel Hill, 1963),
II, pp. 480-1; Deane, *New-England Farmer,* p. 47; Winer, *Great Mountain Forest,* pp. 153-4.

8. That Wilderness Should Turn a Mart

1. "Leift Lion Gardener His Relation of the Pequot Warres," *Massachusetts Historical Society Collections,* 3rd ser., 3 (1833), pp. 154-5.
2. Daniel Gookin, "Historical Collections of the Indians in New England," *Massachusetts Historical Society Collections,* 1st ser., 1 (1792), pp.
142-229.
3. The circumstances of Miantonomo's death can be traced in John
Winthrop, *Winthrop's Journal,* James K. Hosmer, ed. (New York,
1908), II, pp. 131-2, 134-6; Francis Jennings, *The Invasion of America*
(Chapel Hill, 1975), pp. 265-8; Neil Salisbury, *Manitou and Providence*
(New York, 1982), pp. 209-13. A radically different reading of the
event is given by Alden T. Vaughan, *New England Frontier,* rev. ed.
(New York, 1979), pp. 161-6.
4. Maurice Godelier, "The Object and Method of Economic Anthropology" (original French essay first published in 1965), in David
Seddon, ed., *Relations of Production* (London, 1978), p. 61.
5. *Ibid.,* pp. 80-1; Roger Williams, *A Key into the Language of America*
(1643), John J. Teunissen and Evelyn J. Hinz, eds. (Detroit, 1973), p.
124.
6. Marshall Sahlins, *Stone Age Economics* (Chicago, 1972), p. 39; Edward

Johnson, *Johnson's Wonder-Working Providence* (1654), J. Franklin Jameson, ed. (New York, 1910), p. 247.

7. Peter Kalm, *Travels in North America* (1753-61, 1770), Adolph B. Benson, ed. (New York, 1964), I, p. 308; J[oseph] Warren, "Observations on Agriculture," *American Museum*, 2:4 (October 1787), pp. 345, 347. On the importance of this land/labor relationship to economic change in American history, see the seminal essay by H. J. Habbakuk, *American and British Technology in the Nineteenth Century* (Cambridge, England, 1962).

8. Kalm, *Travels*, I, p. 307.

9. Carl O. Sauer, "Theme of Plant and Animal Destruction in Economic History" (1938), in Sauer, *Land and Life* (Berkeley, 1963), p. 154.

BIBLIOGRAPHICAL ESSAY

Because this book has relied so heavily on materials drawn not only from primary documents but also from disciplines whose practitioners are not usually familiar with each other's work, I have tried in this essay to construct a brief guide for those who may wish to pursue further the issues I have discussed. My purpose is to indicate the sources which I myself found most useful, and to suggest the most likely routes of access for those trying to do ecological history in other places and other periods. I have discussed here only the most important materials I have used; readers interested in the details of specific arguments should consult the endnotes.

Primary Documents

Colonial descriptions of the New England landscape break into two broad groups: those written before about 1675, and those written after about 1740. There are surprisingly few materials available of a broadly descriptive nature for the intervening sixty-five years. The most important early accounts, with which any evaluation of colonial ecology must inevitably begin, are William Wood, *New England's Prospect* (1634), Alden T. Vaughan, ed., Amherst, 1977; Thomas Morton, *New English Canaan* (1632), Charles F. Adams, ed., Boston, 1883, which is especially good on Indian interactions with the environment; and John Josselyn's two books, *New-Englands Rarities Discovered* (1672), Edward Tuckerman, ed., *Transactions and Collections of the American Antiquarian*

Society, 4 (1860), pp. 105-238, and *An Account of Two Voyages to New-England* (1675), in *Massachusetts Historical Society Collections*, 3rd ser., 3 (1833), pp. 211-354. Though Josselyn's skills as a naturalist are not entirely reliable, his are among the most thorough seventeenth-century efforts at cataloguing New England plant and animal species.

Even richer than these southern New England sources are the writings of French explorers and missionaries in Nova Scotia, a region whose ecology is similar to that of northern New England. See Pierre Biard's *Relation* (1616) in Reuben Gold Thwaites, ed., *Jesuit Relations, III, Acadia*, Cleveland, 1897; Nicolas Denys, *The Description and Natural History of the Coasts of North America (Acadia)* (1672), William F. Ganong, ed., Toronto, 1908; Chrestien Le Clercq, *New Relation of Gaspesia* (1691), William F. Ganong, ed., Toronto, 1910; and Marc Lescarbot, *The History of New France* (1618), William L. Grant, ed., 3 vols., Toronto, 1907-14. All these works contain extensive details about northern Indian life. Important discussions of the New England coast and its Indian inhabitants prior to European settlement are contained in the earliest explorers' accounts, among which the most important are L. C. Wroth, ed., *The Voyages of Giovanni de Verrazzano, 1524-1528*, New Haven, 1970; H. P. Biggar, ed., *The Works of Samuel de Champlain*, 6 vols., Toronto, 1922-36; Henry S. Burrage, ed., *Early English and French Voyages*, New York, 1906; and Edward Arber, ed., *Travels and Works of Captain John Smith*, Edinburgh, 1910. General secondary reviews of this literature include David B. Quinn, *North America from Earliest Discovery to First Settlements*, New York, 1977; Samuel Eliot Morison, *The European Discovery of America: The Northern Voyages*, New York, 1971; Carl O. Sauer, *Sixteenth-Century North America*, Berkeley, 1971; and Douglas R. McManis, *European Impressions of the New England Coast, 1497-1620*, University of Chicago Geography Department Research Paper No. 139, 1972. One of the best collections of reproductions of early maps is Charles O. Paullin, *Atlas of the Historical Geography of the United States*, Baltimore, 1932.

Documents which speak directly to the settlement of especially southern New England include the two key histories written during the first half of the seventeenth century: William Bradford, *Of Plymouth Plantation*, Samuel Eliot Morison, ed., New York, 1952; and John Winthrop, *Winthrop's Journal*, James K. Hosmer, ed., New York, 1908, both of which give extensive detail about all aspects of early colonial life. Less polished but sometimes at least as suggestive from an ecological point of view is Edward Johnson, *Wonder-Working Providence* (1654), J. Franklin Jameson, ed., New York, 1910. See also the much briefer Francis Higginson, *New-Englands Plantation* (1630), *Massachusetts Historical Society Proceedings*, 62 (1929), pp. 305-21. Documents which supplement these works are conveniently available in Alexander Young, ed., *Chronicles of the Pilgrim Fathers*, Boston, 1841; and Young, ed., *Chronicles of the First Planters of the Colony of Massachusetts Bay*, Boston, 1846. More recent collections of early letters concerning the first settlements are Sydney V. James, Jr., ed., *Three Visitors to Early Plymouth*, Plimoth Plantation, 1963; and Everett Emerson, ed., *Letters from New England*, Amherst, 1976. Samuel Maverick's "A Briefe Discription of New England," *Massachusetts Historical Society Proceedings*, 2nd ser., 1 (1884-5), pp. 231-49, is a valuable account of the New England settlements as they existed in about 1660. Key documents exemplifying the ideology which informed English settlement are the anonymous "Essay on the Ordering of Towns," *Winthrop Papers*, III, 1943, pp. 181-5; and the "Arguments for the Plantation of New England" (sometimes known as "Winthrop's Conclusions"), in the *Winthrop Papers*, II, 1931, pp. 106-49. Those who wish to reconstruct the patterns of English agricultural practice would do well to study Stanley H. Miner and George D. Stanton, eds., *The Diary of Thomas Minor, 1653-1684*, New London, CT, 1899; and *The Diary of Manasseh Minor, 1696-1720*, 1915. John Winthrop, Jr.'s essay on "The Culture and Use of Maize" (1678), reprinted by Fulmer Mood as "John Winthrop, Jr., On Indian Corn," *New England Quarterly*, 10 (1937), pp. 121-33, is a systematic account of maize agriculture as practiced by both Indians and colonists. As regards southern New England Indians, no book is more important than Roger Williams, *A Key into the Language of America* (1643), John J. Teunissen and Evelyn J. Hinz, eds., Detroit, 1973. This should be supplemented by *The Letters of Roger Williams*, John R. Bartlett, ed., Providence, 1874. Also extremely valuable is Daniel Gookin, "Historical Collections of the Indians in New England," *Massachusetts Historical Society Collections*, 1st ser., 1 (1792), pp. 141-227, which was written just prior to King Philip's War.

As I note in the first chapter, legal records can reveal much about a variety of practices which affected the colonial environment. The key colony-wide collections are Nathaniel B. Shurtleff, ed., *Records of the Governor and Company of the Massachusetts Bay in New England*, Boston, 1853; Shurtleff, ed., *Records of the Colony of New Plymouth*, Boston, 1855; Charles J. Hoadly, ed., *The Public Records of the Colony of Connecticut*, Hartford, 1850-90; John R. Bartlett, ed., *Records of the Colony of Rhode Island and Providence Plantation*, Providence, 1856; and Charles J. Hoadly, ed., *Records of the Colony or Jurisdiction of New Haven*, Hartford, 1858. Town records should be examined as well, but are too numerous to be listed here; the best bibliography I know which lists published town records is in Edward M. Cook, Jr., *The Fathers of the Towns*, Baltimore, 1976, pp. 237-65. Researchers should also note the helpful topical compilations of colonial laws in *Laws of the Colonial and State Governments Relating to the Indians and Indian Affairs*, Washington, 1832; and J. P. Kinney, "Forest Legislation in America Prior to March 4, 1789," *Cornell University Agricultural Experiment Station Bulletin*, 370 (1916), pp. 357-405.

Not until the late eighteenth century do we get extensive writings by American observers who describe the ecological changes going on around them. By far the most valuable of these is Timothy Dwight's *Travels in New England and New York* (1821), 4 vols., Barbara Miller Solomon, ed., Cambridge, MA, 1969; anyone interested in New England ecology could do no better than to read Dwight from cover to cover. Also very important are Peter Kalm's *Travels in North America* (1753-61, 1770), Adolph B. Benson, ed., 2 vols., New York, 1964, which although it deals with the mid-Atlantic colonies has very shrewd observations that can often be generalized to New England; Peter Whitney, *A History of the County of Worcester*, Worcester, 1793, which has extensive notes on the topography of the county's towns; and Jeremy Belknap's superb third volume to his *History of New Hampshire*, Dover, NH, 1812. A number of European travelers' accounts contain suggestive fragments about the New England environment; for references to these, readers should see my notes. On agriculture in the second half of the eighteenth century, three books stand out: Jared Eliot, *Essays upon Field Husbandry in New England and Other Papers, 1748-1762*, Harry J. Carman and Rexford G. Tugwell, eds., New York, 1934; Harry J. Carman, ed., *American Husbandry* (1775), New York, 1939; and Samuel Deane, *The New-England Farmer*, Worcester, MA, 1790. General James Warren's "Observations on Agriculture," *American Museum*, 2:4 (October 1787), pp. 344-8, is also well worth examining for its comparison of British and American agriculture. Finally, a unique document by a southern New England Indian is the "Extract from an Indian History," *Massachusetts Historical Society Collections,* 1st ser., 9 (1804), pp. 99-102.

Ecological Literature

Few of the sources listed above, obviously, adopt an explicitly ecological perspective on the places and economic practices they describe. (Readers wishing to investigate the state of colonial science and natural history might begin by consulting George Browne Goode, "The Beginnings of Natural History in America," *Proceedings of the Biological Society of Washington,* 3 [1886], pp. 35-105; Henry Savage, *Lost Heritage,* New York, 1970; W. M. and Mabel Smallwood, *Natural History and the American Mind,* New York, 1941; and Raymond P. Stearns, *Science in the British Colonies of America,* Urbana, 1970.) The best way for a modern historian to bring such a perspective to the documents is to get out and walk the landscape: no amount of library work can replace the field experience gained by exploring different habitats as they exist today. Even though such habitats are usually altered from their earlier conditions, learning to perceive ecological relationships within them is essential if a historian is to try to reconstruct past environments. A variety of field guides are available to help non-ecologists gain access to different plant and animal communities; it is to these that readers should turn if they are confused by my use of different species names in the body of the text. Two guides by Neil Jorgensen are superb on the overall ecology of New England: *A Sierra Club Naturalist's Guide to Southern New England,* San Francisco, 1978; and *A Guide to New England's Landscape,* Chester, CT, 1977. Equally good on coastal habitats is Michael and Deborah Berrill's *A Sierra Club Naturalist's Guide to the North Atlantic Coast,* San Francisco, 1981. On tree species, which any would-be ecological historian would do well to know, Elbert L. Little's *The Audubon Society Field Guide to North American Trees: Eastern Region,* New York, 1980, is beautifully illustrated and easy to use; George A. Petrides, *A Field Guide to Trees and Shrubs,* 2nd ed., Boston, 1972, in the Peterson series, includes shrubs and is thus more complete, but its taxonomic keys may be harder for the inexperienced to use. Other field guides to flowering plants, birds, mammals, insects, and so on, are widely available in the Peterson, Golden, and Audubon series; readers should judge for themselves which of these will work best for them.

There are an increasing number of good textbooks on modern ecological theory. The standard volume for many years has been Eugene P. Odum's *Fundamentals of Ecology*, 3rd ed., Philadelphia, 1971, which is well written and encyclopedic, though now somewhat dated. Robert L. Smith, *Ecology and Field Biology*, 3rd ed., New York, 1980, has a more habitat-oriented approach that may make it more accessible to lay readers. Two older texts that are still excellent as brief introductions to the field are Eugene P. Odum, *Ecology*, New York, 1963; and Edward J. Kormondy, *Concepts of Ecology*, Englewood Cliffs, NJ, 1969. Texts which incorporate important recent developments from animal population studies include Robert E. Ricklefs, *Ecology*, 2nd ed., New York, 1979; Paul A. Colinvaux, *Introduction to Ecology*, New York, 1973; J. Merritt Emlen, *Ecology: An Evolutionary Approach*, Menlo Park, 1973; Charles J. Krebs, *Ecology: The Experimental Analysis of Distribution and Abundance*, New York, 1972; Boyd D. Collier, *et al.*, *Dynamic Ecology*, Englewood Cliffs, NJ, 1973; and G. Evelyn Hutchinson, *An Introduction to Population Ecology*, New Haven, 1978. A text which emphasizes human alteration of natural ecosystems in a wide-ranging (if sometimes polemical) fashion is Paul R. Ehrlich, *et al.*, *Ecoscience*, San Francisco, 1977; historians may find it particularly suggestive. More specialized discussions which I found helpful in constructing my own analysis were Stephen H. Spurr and Burton V. Barnes, *Forest Ecology*, 2nd ed., New York, 1973; Joseph Kittredge, *Forest Influences*, New York, 1948; and Richard Lee, *Forest Hydrology*, New York, 1980. Much of my understanding of forest nutrient export comes from the Hubbard Brook studies described by F. Herbert Bormann and Gene E. Likens in *Pattern and Process in a Forested Ecosystem*, New York, 1979. See also their more accessible "Catastrophic Disturbance and the Steady State in Northern Hardwood Forests," *American Scientist*, 67 (1979), pp. 660-9; and Bormann, *et al.*, "The Export of Nutrients and Recovery of Stable Conditions Following Deforestation at Hubbard Brook," *Ecological Monographs*, 44 (1974), pp. 255-77. On wildlife populations, see the excellent popular history by Peter Matthiessen, *Wildlife in America*, New York, 1959; A. W. Schorger, *The Passenger Pigeon*, Madison, 1955; and Schorger, *The Wild Turkey*, Norman, OK, 1966. Good general introductions to the techniques used in reconstructing past environments are Karl W. Butzer, *Environment and Archeology*, 2nd ed., Chicago, 1971; Butzer, *Archaeology as Human Ecology*, Cambridge, England, 1982; and John G. Evans, *An Introduction to Environmental Archaeology*, Ithaca, 1978. Historians interested in the intellectual history of ecology as a science should consult Donald Worster, *Nature's Economy*, San Francisco, 1977; and Ronald C. Tobey, *Saving the Prairies: The Life Cycle of the Founding School of American Plant Ecology, 1895-1955*, Berkeley, 1981. Those who wish to keep track of current literature in the field should at a minimum consult the journals *Ecology*, *Ecological Monographs*, and the *Annual Review of Ecology and Systematics*.

Those seeking good general accounts of New England ecosystems might begin with Betty Flanders Thomson's accessible and delightfully written *The Changing Face of New England*, Boston, 1958; an equally competent popular account of salt marsh ecology is John and Mildred Teal, *Life and Death of the Salt Marsh*, New York, 1969. Briefer but more technical descriptions are Douglas S. Byers, "The Environment of the Northeast," in Frederick Johnson, ed., *Man in Northeastern North America*, Papers of the Robert S. Peabody Foundation for Archaeology, 3 (1946), pp. 3-32; and John W. Barrett, "The Northeastern Region," in Barrett, ed., *Regional Silviculture of the United States*, 2nd ed., New York, 1980, pp. 25-65. Much more technical but encyclopedic in its coverage is E. Lucy Braun, *Deciduous Forests of Eastern North America*, New York, 1950; see also H. A. Gleason, *Plants in the Vicinity of New York*, rev. ed., New York, 1962. An older and more outdated source which historians may nevertheless find useful is George B. Emerson, *A Report on the Trees and Shrubs Growing Naturally in the Forests of Massachusetts*, Boston, 1846. The maps in Howard W. Lull, *A Forest Atlas of the Northeast*, Upper Darby, PA, 1968, will be helpful to those trying to visualize regional environmental patterns.

One way ecologists have sought to gain access to precolonial vegetational communities is to examine stands of old-growth timber. There are problems in doing this: virtually no uncut forests survive today, so that all existing stands are at least "second-growth," and even the oldest of these have potentially been modified by a variety of human activities. (Still more troubling is the question of whether or not a very old stand of timber as it exists today accurately represents the forest mosaic of different successional stages which Indians inhabited and modified.) Studies of old forests can nevertheless be quite suggestive. For examples, see G. E. Nichols, "The Vegetation of Connecticut, II, Virgin Forests," *Torreya*, 13 (1913), pp. 199-215; H. J. Lutz, "The Vegetation of Heart's Content, A Virgin Forest in Northwestern Pennsylvania," *Ecology*, 11 (1930), pp. 1-29; Hugh M. Raup, "An Old Forest in Stonington, Connecticut," *Rhodora*, 43 (1941), pp. 67-71; A. C. Cline and S. H. Spurr, "The Virgin Upland Forest of Central New England: A Study of Old Growth Stands in the Pisgah Mountain Section of Southwestern New Hampshire," *Harvard Forest Bulletin*, 21 (1942); and F. H. Bormann and M. F. Buell, "Old Age Stand of Hemlock-Northern Hardwood Forest in Central Vermont," *Bulletin of the Torrey Botanical Club*, 91 (1964), pp. 451-65. Studies which attempt more general reconstructions of associations between tree species in the forest habitats of different New England regions include G. E. Nichols, "The Hemlock-White Pine-Northern Hardwood Region of Eastern North America," *Ecology*, 16 (1935), pp. 403-22; S. H. Spurr, "Forest Association in the Harvard Forest," *Ecological Monographs*, 26 (1956), pp. 245-62; and Ronald B. Davis, "Spruce-Fir Forests of the Coast of Maine," *Ecological Monographs*, 36 (1966), pp. 79-94. H. J. Lutz, "Trends and Silvicultural Significance of Upland Forest Successions in Southern New England," *Yale University School of Forestry Bulletin*, 22 (1928), is an excellent summary not only of associations, but of successional sequences on agricultural and pastured lands. A fine paper on the structure and evolution of New England salt marshes is Alfred C. Redfield, "Development of a New England Salt Marsh," *Ecological Monographs*, 42 (1972), pp. 201-37.

Pollen studies are another method ecologists have used to reconstruct precolonial environments. A standard textbook on palynology is K. Faegri and J. Iversen, *Textbook of Pollen Analysis*, 3rd ed., Copenhagen, 1975; readers seeking a less daunting introduction might try consulting Margaret B. Davis, "Palynology and Environmental History During the Quaternary Period," *American Scientist*, 57 (1969), pp. 317-32, which uses New England examples, or her "On the Theory of Pollen Analysis," *American*

Journal of Science, 261 (1963), pp. 897-912. Pollen analysis has proved to be most useful in inferring the character of long-term vegetational and climatic shifts in the postglacial period: useful syntheses of this material can be found in Margaret B. Davis, "Phytogeography and Palynology of Northeastern United States," in H. E. Wright and David G. Frey, eds., *The Quaternary of the United States,* Princeton, 1965, pp. 377-401 (an excellent article); H. E. Wright, "Late Quaternary Vegetational History of North America," in Karl K. Turekian, ed., *Late Cenozoic Glacial Ages,* New Haven, 1971, pp. 425-64, which covers most of North America; and Thompson Webb III, "The Past 11,000 Years of Vegetational Change in Eastern North America," *Bioscience,* 31 (1981), pp. 501-6, which contains an excellent series of climatic maps. (The French historian Emmanuel Le Roy Ladurie has used pollen data in conjunction with a wide variety of other sources to reconstruct European climates since A.D. 1000 in his well-known *Times of Feast, Times of Famine,* New York, 1971.) Unfortunately, the very success of American pollen scientists in analyzing climates of the relatively distant past has led them until recently to devote little attention to changes in pollen composition following the European arrival in North America. The advent of radiocarbon dating has now made studies of the post-European period more feasible. The most important of these for New England is R. B. Brugam, *The Human Disturbance History of Linsley Pond, North Branford, Connecticut,* Ph.D. Thesis, Yale University, 1975, which is summarized in part in Brugam's "Pollen Indicators of Land-Use Change in Southern Connecticut," *Quaternary Research,* 9 (1978), pp. 349-62; see also Emily W. Russell, *Vegetational Change in Northern New Jersey since 1500 A.D.: A Palynological, Vegetational, and Historical Synthesis,* Ph.D. Thesis, Rutgers University, 1979. Other studies which devote some attention to the influence of human beings on pollen and sediment deposition rates include Margaret B. Davis, "Pollen Evidence of Changing Land Use around the Shores of Lake Washington," *Northwest Science,* 47 (1973), pp. 133-48; her "Erosion Rates and Land Use History in Southern Michigan," *Environmental Conservation,* 3 (1976), pp. 139-48; and Thompson Webb III, "A Comparison of Modern and Presettlement Pollen from Southern Michigan," *Review of Palaeobotany and Palynology,* 16 (1973), pp. 137-56.

Historians will probably gain the most by reading those studies in which ecologists have tried to analyze the effects of human activity on forests and other natural communities. A brief survey of this literature can be found in Stephen Spurr's "The American Forest Since 1600," in his *Forest Ecology*, pp. 475-93. Two classic articles which must not be overlooked are Stanley W. Bromley, "The Original Forest Types of Southern New England," *Ecological Monographs*, 5 (1935), pp. 61-89; and Gordon M. Day, "The Indian as an Ecological Factor in the Northeastern Forest," *Ecology*, 34 (1953), pp. 329-46, both of which deal at length with the effects of fire on forest habitats. The literature on fire ecology is large; those seeking nontechnical introductions to it should see Charles F. Cooper, "The Ecology of Fire," *Scientific American*, 204:4 (April 1961), pp. 150-60; and D. Q. Thompson and R. H. Smith, "The Forest Primeval in the Northeast—A Great Myth?" in *Proceedings of the Annual Tall Timbers Fire Ecology Conference*, 10 (1970), pp. 255-65. Richer but more technical are Silas Little, "Effects of Fire on Temperate Forests: Northeastern United States," in T. T. Kozlowski and C. E. Ahlgren, eds., *Fire and Ecosystems*, New York, 1974, pp. 225-50; William A. Niering, *et al.*, "Prescribed Burning in Southern New England," *Proceedings of the Annual Tall Timbers Fire Ecology Conference*, 10 (1970), pp. 267-86; and all of the articles in the superb October 1973 (3:3) issue of *Quaternary Research*, especially the fine introduction by H. E. Wright and M. L. Heinselman, pp. 319-28. An excellent recent history of fire is Stephen J. Pyne, *Fire in America*, Princeton, 1982.

Several older ecological studies attempt to write the history of particular forests in New England. These include Hugh M. Raup and Reynold E. Carlson, "The History of Land Use in the Harvard Forest," *Harvard Forest Bulletin*, 20 (1941); Raup's more popular "The View from John Sanderson's Farm," *Forest History*, 10 (1966), pp. 2-11; J. Wilcox Brown, "Forest History of Mt. Moosilauke," *Appalachia*, 24 (1958), pp. 23-32, 221-33; H. I. Winer, *History of the Great Mountain Forest, Litchfield County, Connecticut*, Ph.D. Thesis, Yale University, 1955; and J. G. Ogden, "Forest History of Martha's Vineyard: I. Modern and Pre-Colonial Forests," *American Midland Naturalist*, 66 (1961), pp. 417-30. Three more recent studies which make fascinating use of the decaying plant materials in forest floors to reconstruct stand histories are C. D. Chadwick and E. P. Stephens, "Reconstruction of a Mixed-Species Forest in Central New England," *Ecology*, 58 (1977), pp. 562-72;

J. D. Henry and J. M. A. Swan, "Reconstructing Forest History from Live and Dead Plant Material," *Ecology*, 55 (1974), pp. 772-83; and Rebecca Ellen Bormann, *Agricultural Disturbance and Forest Recovery at Mt. Cilley*, Ph.D. Thesis, Yale University, 1982. Ecologists have long been using the federal land survey records to map out precolonial vegetation patterns, but of course such data are available only in northern New England; the only New England studies to make use of such records are Thomas G. Siccama, "Presettlement and Present Forest Vegetation in Northern Vermont with Special Reference to Chittenden County," *American Midland Naturalist*, 85 (1971), pp. 153-72; and Craig G. Lorimer, "The Presettlement Forest and Natural Disturbance Cycle of Northeastern Maine," *Ecology*, 58 (1977), pp. 139-48. Historians working in the trans-Appalachian West should be aware that a large literature exists using land survey data, and might at least familiarize themselves with these techniques by reading E. A. Bourdo's classic "A Review of the General Land Office Survey and of Its Use in Quantitative Studies of Former Forests," *Ecology*, 37 (1956), pp. 754-68.

Finally, a handful of non-New England case studies, written principally by historians and geographers, should be mentioned as potential models for future efforts at writing ecological history. An extremely rich volume containing examples from around the world should be among the first books consulted by anyone interested in this subject: William L. Thomas, ed., *Man's Role in Changing the Face of the Earth*, Chicago, 1956. A more recent volume that is similarly comprehensive is Andrew Goudie, *The Human Impact: Man's Role in Environmental Change*, Cambridge, MA, 1981. Also worth consulting are George Perkins Marsh's classic *Man and Nature* (1864), David Lowenthal, ed., Cambridge, MA, 1965; and Lucien Febvre, *A Geographical Introduction to History*, New York, 1925. Nearly all the writings of Carl O. Sauer are valuable, but probably the most important are *The Early Spanish Main*, Berkeley, 1966; and the two collections of essays, *Land and Life*, Berkeley, 1963; and its companion, the misleadingly titled *Selected Essays, 1963-1975*, Berkeley, 1981. A contemporary of Sauer's whose work has been undeservedly neglected is James C. Malin; see his *Winter Wheat in the Golden Belt of Kansas*, Lawrence, KS, 1944; and *The Grassland of North America*, Lawrence, KS, 1947. Four historians have recently written works which make significant contributions to ecological history. Alfred W. Crosby's *The Columbian Exchange: Biological and Cultural Consequences of 1492,* Westport, CT, 1972, is a thorough analysis of the exchange of species between the Old World and the New, with particular emphasis on diseases. Readers can get a quick introduction to his approach in his "Ecological Imperialism: The Overseas Migration of Western Europeans as a Biological Phenomenon," *Texas Quarterly,* 80 (1978), pp. 10-22. William H. McNeill has pursued a similar thesis on a global scale, analyzing the movement of disease organisms among human communities around the world: see his *Plagues and Peoples,* New York, 1976; and *The Human Condition,* Princeton, 1980. E. L. Jones has sought to explain the development of modern Europe by using concepts drawn from ecological population theory in his *The European Miracle,* Cambridge, England, 1981; American readers will also be interested in his "Creative Disruptions in American Agriculture, 1620-1820," *Agricultural History,* 48 (1974), pp. 510-28. Calvin Martin has offered an ecological interpretation of North American Indian life in his *Keepers of the Game,* Berkeley, 1978; in addition to the criticisms of it I offer in my text, readers should consult Shepard Krech III, ed., *Indians, Animals and the Fur Trade: A Critique of Keepers of the Game,* Athens, GA, 1981. Six works which choose smaller geographical units of analysis for their histories of ecological change are Andrew Hill Clark, *The Invasion of New Zealand by People, Plants and Animals,* New Brunswick, NJ, 1949; David Watts, "Man's Influence on the Vegetation of Barbados, 1627-1800," *University of Hull Occasional Papers in Geography,* 4 (1966); John W. Bennett, *Northern Plainsmen,* Chicago, 1969; Donald Worster, *Dust Bowl,* New York, 1980; Richard White, *Land Use, Environment, and Social Change: The Shaping of Island County, Washington,* Seattle, 1980; and William L. Preston, *Vanishing Landscapes: Land and Life in the Tulare Lake Basin,* Berkeley, 1981. White's book in particular seems to me a model for future work in this field. There is a large body of literature written by European historians which bears on the problem of doing ecological history, but I cannot survey it here.

Ecological and Economic Anthropology

Anthropologists have engaged in extensive discussions of how ecological theory should be incorporated into the study of human populations. The literature here is quite large, but non-anthropologists can survey it in one of four recent textbooks: John W. Bennett's theoretical essay, *The Ecological Transition: Cultural Anthropology and Human Adaptation*, New York, 1976; Robert McC. Netting's brief examination of case studies in *Cultural Ecology*, Menlo Park, 1977; Donald Hardesty's general survey text, *Ecological Anthropology*, New York, 1977; and Emilio Moran, *Human Adaptability: An Introduction to Ecological Anthropology*, North Scituate, 1979. Hardesty in particular has an extensive bibliography. Those who wish a more technical review of the literature can turn to any of a number of bibliographical essays that have appeared in the last two decades. The most extensive is probably James N. Anderson, "Ecological Anthropology and Anthropological Ecology," in John J. Honigmann, ed., *Handbook of Social and Cultural Anthropology*, Chicago, 1973, pp. 179-239, although this is now quite dated. More recent are Robert McC. Netting, "Agrarian Ecology," *Annual Review of Anthropology*, 3 (1974), pp. 21-56; Andrew P. Vayda and Bonnie J. McCay, "New Directions in Ecology and Ecological Anthropology," *Annual Review of Anthropology*, 4 (1975), pp. 293-306; and Benjamin S. Orlove, "Ecological Anthropology," *Annual Review of Anthropology*, 9 (1980), pp. 235-73. Older essays that are still worth examining include Marston Bates, "Human Ecology," in A. L. Kroeber, ed., *Anthropology Today*, Chicago, 1953, pp. 700-13; Julian Steward's seminal *Theory of Culture Change*, Urbana, 1955, esp. pp. 30-42; June Helm, "The Ecological Approach in Anthropology," *American Journal of Sociology*, 67 (1962), pp. 630-9; Marshall D. Sahlins, "Culture and Environment: The Study of Cultural Ecology," in Robert A. Manners and Donald Kaplan, eds., *Theory in Anthropology*, Chicago, 1968, pp. 367-73; Andrew P. Vayda and Roy A. Rappaport, "Ecology, Cultural and Noncultural," in James A. Clifton, ed., *Introduction to Cultural Anthropology*, Boston, 1968, pp. 477-97; and Roy A. Rappaport, "Nature, Culture, and Ecological Anthropology," in Harry L. Shapiro, ed., *Man, Culture and Society*, New York, 1971, pp. 237-67. A useful collection of readings in the field is Andrew P. Vayda, ed., *Environment and Cultural Behavior*, Garden City, NY, 1969.

The other major subfield of anthropology which examines human interactions with the environment is economic anthropology. Relations and systems of production in human communities inevitably entail manipulation of surrounding environments, and our best point of departure for explaining why different peoples have different effects on an ecosystem is to examine their respective economies. Economic anthropology has been split since the mid-1950s between so-called formalists and substantivists. The former believe that the abstract, market-oriented principles of neoclassical economics can be fruitfully applied to most non-Western societies; the latter reject this as an ahistorical claim, arguing that each society possesses a more or less unique economic logic which must be considered on its own conceptual terms. The classic formalist textbook is Melville J. Herskovits, *Economic Anthropology*, New York, 1952. The substantivist critique was first articulated in the now famous volume edited by Karl Polanyi, Conrad Arensberg, and Harry Pearson, *Trade and Market in Early Empires*, New York, 1957. Polanyi is the leading figure of the school, and his emphasis on the economy as an instituted process is one that some ecological anthropologists have found fruitful. His essays have been collected in *Primitive, Archaic, and Modern Economies*, George Dalton, ed., Boston, 1968; and S. C. Humphreys has evaluated his contribution in "History, Economics, and Anthropology: The Work of Karl Polanyi," *History and Theory*, 8 (1969), pp. 165-212. Polanyi's chief disciple is George Dalton, whose work can be sampled in *Economic Anthropology and Development*, New York, 1971; and "The Impact of Colonization on Aboriginal Economies in Stateless Societies," in Dalton, ed., *Research in Economic Anthropology*, Greenwich, CT, 1 (1978), pp. 131-84, an essay that is particularly relevant to this book. Dalton has also edited a useful collection of articles that are primarily substantivist in their orientation: *Tribal and Peasant Economies*, Garden City, NY, 1967. This should be compared with the essays in Raymond Firth, ed., *Themes in Economic Anthropology*, London, 1967.

The substantivists, for all of their cogency in pointing out the absurdity of too simple a transfer of Western economic concepts to non-Western societies, have been criticized, I think rightly, for throwing the baby out with the bathwater, denying even the possibility of a theoretical framework for cross-cultural comparisons of political economy. One of the early critics to point this out was Scott Cook, "The Obsolete 'Anti-Market' Mentality: A Critique of the Substantivist Approach to Economic Anthropology," *American Anthropologist*, 68 (1966), pp. 323-45; see also Edward

E. LeClair, Jr., "Economic Theory and Economic Anthropology," *American Anthropologist*, 64 (1962), pp. 1179-1203. The substantivist counterattack was David Kaplan's "The Formal-Substantive Controversy in Economic Anthropology," *Southwestern Journal of Anthropology*, 24 (1968), pp. 228-51; to which Cook replied in "The 'Anti-Market' Mentality Reexamined," *Southwestern Journal of Anthropology*, 25 (1969), pp. 378-406. By the early 1970s, it was clear to many that the debate was becoming sterile, and Cook's review essay, "Economic Anthropology," in John J. Honigmann, ed., *Handbook of Social and Cultural Anthropology*, Chicago, 1973, pp. 795-860 is an excellent survey of the literature up to that time. Efforts at synthesis have tended to look to ecology and to the structuralism of the French Marxist anthropologists for possible ways of integrating the two positions. Marshall Sahlins, "Economic Anthropology and Anthropological Economics," *Social Science Information*, 8:5 (1969), pp. 13-33 made early suggestions about the utility of ecological formulations, and Scott Cook's "Production, Ecology, and Economic Anthropology," *Social Science Information*, 12:1 (1973), pp. 25-52, made these still more explicit. Sahlins's work has been very rich in this respect: see his *Tribesmen*, Englewood Cliffs, NJ, 1968, for a general survey that is helpful for fitting the New England Indians into a broader context; and his seminal *Stone Age Economics*, Chicago, 1972, which should be supplemented with the valuable collection by Richard B. Lee and Irven Devore, eds., *Man the Hunter*, New York, 1968. Ester Boserup's much criticized *The Conditions of Agricultural Growth*, Chciago, 1965, can also be quite fruitful for those trying to assess the ecological effects of non-Western societies. Those interested in examining recent Marxist work in these areas should begin with the very useful collection edited by David Seddon, *Relations of Production: Marxist Approaches to Economic Anthropology*, London, 1978. Marx's own work on noncapitalist societies is conveniently collected in *Pre-Capitalist Economic Formations*, Eric J. Hobsbawm, ed., New York, 1964; Marx and Engels, *The German Ideology*, New York, 1970, is also worth examining. Seddon's collection should be followed by a reading of Maurice Godelier's *Rationality and Irrationality in Economics*, New York, 1972; and *Perspectives in Marxist Anthropology*, Cambridge, England, 1977. A brilliant if abrasive Marxist critique of an overly functionalist ecological anthropology is Jonathan Friedman's "Marxism,

Structuralism and Vulgar Materialism," *Man*, n.s., 9 (1974), pp.
444-69; this should be read in conjunction with Roy Rappaport's
response, "Ecology, Adaptation and the Ills of Functionalism,"
Michigan Discussions in Anthropology, 2 (Winter 1977), pp. 138-90.
The collection edited by Philip C. Burnham and Roy F. Ellen,
Social and Ecological Systems, New York, 1979, is valuable in suggesting possible lines of synthesis.

The New England Indians

The starting point for any research on New England Indians
must be the superb fifteenth volume of the new *Handbook of North
American Indians* (Washington, 1978), which is entitled *Northeast*
and edited by Bruce Trigger. Its eighty-three-page bibliography
is comprehensive. Also of use in surveying the literature is Eli-
sabeth Tooker, *The Indians of the Northeast: A Critical Bibliography*,
Bloomington, 1978. The continent-wide *Ethnographic Bibliography
of North America*, edited by George Peter Murdock and Timothy
J. O'Leary, 4th ed., 5 vols., New Haven, 1975, is often helpful.
Recent review essays which survey the historical literature are
Bernard Sheehan, "Indian-White Relations in Early America,"
William and Mary Quarterly, 3rd ser., 26 (1969), pp. 267-86; Francis
Jennings, "Virgin Land and Savage People," *American Quarterly*,
23 (1971), pp. 519-41; and James Axtell, "The Ethnohistory of Early
North America," *William and Mary Quarterly*, 3rd ser., 35 (1978),
pp. 110-44. An interesting collection of essays about the environ-
mental relationships of North American Indians generally is
Christopher Vecsey and Robert W. Venables, eds., *American Indian Environments*, Syracuse, 1980.

Several works perform the very useful function of collecting
and, in effect, indexing the available primary documents in order
to depict some aspect of New England life. Charles C. Wil-
loughby surveys New England archaeological objects in his *An-
tiquities of the New England Indians*, Cambridge, MA, 1935. Froehlich G. Rainey collates most of the major primary sources in his
helpful "A Compilation of Historical Data Contributing to the
Ethnography of Connecticut and Southern New England Indians," *Bulletin of the Archaeological Society of Connecticut*, 3 (April
1936), pp. 1-89. Regina Flannery performs the same function for

the entire East Coast, albeit in a more schematic format, in her "An Analysis of Coastal Algonquin Culture," *Catholic University of America Anthropological Series*, 7 (1939). Catherine Marten's "The Wampanoags in the Seventeenth Century: An Ethno-Historical Survey," *Occasional Papers in Old Colony Studies*, 2 (1970), pp. 1-40, is thorough in its survey of the early evidence from the Massachusetts Bay area.

The most comprehensive studies yet published of seventeenth-century New England Indians are Alden T. Vaughan, *New England Frontier: Puritans and Indians, 1620-1675*, rev. ed., New York, 1979, which draws a picture so partial to the colonists as to be almost an apology for them; and Francis Jennings, *The Invasion of America: Indians, Colonialism, and the Cant of Conquest*, Chapel Hill, 1975, which sometimes argues dangerously from negative evidence and perhaps goes too far in the opposite direction in its polemic against colonial injustices, but basically gets the story straight. Between the two, I'd choose Jennings. T. J. C. Brasser's "The Coastal Algonkians," in Eleanor Leacock and Nancy Lurie, eds., *North American Indians in Historical Perspective*, New York, 1971, pp. 64-91, is a good brief survey of all the coastal Algonquians. Three more recent books concentrate on narrower themes. Karen Ordahl Kupperman, *Settling with the Indians*, Totowa, NJ, 1980, is useful in pointing to the *similarities* rather than the *differences* between Indians and colonists, similarities which my own account tends not to emphasize. (Kupperman basically elaborates a point made about the Virginia Indians by Nancy Lurie in her classic essay "Indian Cultural Adjustment to European Civilization," in James M. Smith, ed., *Seventeenth-Century America*, Chapel Hill, 1959, pp. 33-60. P. Richard Metcalf's "Who Should Rule at Home? Native American Politics and Indian-White Relations," *Journal of American History*, 61 [1974], pp. 651-65, makes a similar argument but seems to me to overlook structural economic differences between Indian and European societies.) James Axtell's collected essays in *The European and the Indian*, New York, 1981, deal less with economic and ecological relationships between Indians and colonists than with religious and cultural ones, but are nevertheless essential reading. Neil Salisbury, *Manitou and Providence*, New York, 1982, is extremely detailed in his readings of Indian-colonial interaction in the first four decades of the seventeenth century, and pays some attention to ecological questions. Those interested in northern hunting peoples may wish to consult Frank G. Speck, *Penobscot Man*, Philadelphia, 1940, which is based more on ethnographic than on historical sources; Alfred Goldsworthy Bailey, *The Conflict of European and Eastern Algonkian Cultures, 1504-1700*, 2nd ed., Toronto, 1969; Kenneth M. Morrison, *The People of the Dawn: The Abnaki and Their Relations with New England and New France, 1600-1727*, Ph.D. Thesis, University of Maine, 1975; and Cornelius J. Jaenen, *Friend and Foe: Aspects of French-Amerindian Cultural Contact in the Sixteenth and Seventeenth Centuries*, Toronto, 1976. Calvin Martin, *Keepers of the Game*, deals with the same groups.

Archaeologists have been among the most perceptive analysts of interactions between New England Indians and their environments. Dean R. Snow's *The Archaeology of New England*, New York, 1980, is the best way of gaining access to this literature, and is a superb synthesis. William A. Haviland and Marjory W. Power, *The Original Vermonters*, Hanover, NH, 1981, is limited to a single state but is a full-scale scholarly synthesis as well. Susan Gibson, ed., *Burr's Hill*, Providence, 1980, and William C. Simmons, *Cautantowwit's House*, Providence, 1970, are well-illustrated and well-written records of the excavations of Indian burial grounds from the postcontact period, and are especially useful in their discussions of European trade goods. Two doctoral dissertations use archaeological data to examine changes in Indian settlement patterns on Long Island Sound as a result of Indian-European interaction: Lorraine Williams, *Ft. Shantok and Ft. Corchaug: A Study of Seventeenth Century Culture Contact in the Long Island Sound Area*, Ph.D. Thesis, New York University, 1972; and Lynn Ceci, *The Effect of European Contact and Trade on the Settlement Pattern of Indians in Coastal New York, 1524-1665*, Ph.D. Thesis, City University of New York, 1977. Finally, Peter A. Thomas's doctoral dissertation stands in a class by itself as a very sophisticated assessment of the different ecological relationships of Indians and colonists: *In the Maelstrom of Change: The Indian Trade and Cultural Process in the Middle Connecticut River Valley, 1635-1665*, Ph.D. Thesis, University of Massachusetts, 1979. Anyone wishing to pursue ecological history in New England should regard this thesis as essential reading, but a quick summary of its argument can be obtained in Thomas's "Contrastive Subsistence Strategies and Land Use as Factors for Understanding Indian-White Relations in New England," *Ethnohistory*, 23 (1976), pp. 1-18.

A variety of good studies examine the material culture and economies of New England Indians. Howard S. Russell, *Indian New England Before the Mayflower*, Hanover, NH, 1980, although basically antiquarian and uninformed by anthropological theory, is very thorough in its coverage of all aspects of Indian material life; the book's bibliography is undigested but extraordinarily extensive. Two doctoral dissertations are especially fine for their ethnographic discussions: Robert Austin Warner, *The Southern New England Indians to 1725: A Study in Culture Contact*, Ph.D. Thesis, Yale University, 1935, deals with the agricultural peoples of the south; and Bernard Hoffman is excellent on the ecological relationships of northern Indians in his *Historical Ethnography of the Micmac of the Sixteenth and Seventeenth Centuries*, Ph.D. Thesis, UCLA, 1955. See also his "Ancient Tribes Revisited: A Summary of Indian Distribution and Movement in the Northeastern United States from 1534 to 1779," *Ethnohistory*, 14 (1967), pp. 1-46. James Axtell's *Indian Peoples of Eastern North America*, New York, 1981, is a useful collection of primary documents dealing mainly with gender relationships in Indian communities. Several articles analyze Indian diets. Eva L. Butler, "Algonkian Culture and Use of Maize in Southern New England," *Bulletin of the Archaeological Society of Connecticut*, 22 (December 1948), pp. 2-39, is quite thorough and quotes extensively from the primary sources. William S. Fowler, "Agricultural Tools and Techniques of the Northeast," *Massachusetts Archaeological Society Bulletin*, 15 (1954), pp. 41-51, is mainly useful for its archaeological illustrations of agricultural tools. M. K. Bennett's "The Food Economy of the New England Indians, 1605-1675," *Journal of Political Economy*, 63 (1955), pp. 369-97, is an economist's effort to reconstruct the caloric content of Indian diets but has numerous statistical problems which I discuss in my endnotes. Frederic W. Warner, "The Foods of the Connecticut Indians," *Bulletin of the Archaeological Society of Connecticut*, 37 (1972), pp. 27-47, combines historical and archaeological data in a competent analysis of the whole range of foodstuffs used by southern New England Indians. On Indian agriculture, Lynn Ceci, "Fish Fertilizer: A Native North American Practice?" *Science*, 188 (1975), pp. 26-30, caused quite a stir by denying that New England Indians had used fertilizer; replies to her article are in *Science*, 189 (1975), pp. 944-50. Readers should compare my account of shifting agriculture in New England with anthropological descriptions of swidden agriculture in the tropics. The literature is enormous, but the classic articles are by Harold C. Conklin, "An Ethnocultural Approach to Shifting Agriculture," *Transactions of the New York Academy of Science*, Series II, 17 (1954), pp. 133-42; and "The Study of Shifting Cultivation," *Current Anthropology*, 2 (1961), pp. 27-61. Indian burning practices can be examined in the previously cited article by Day; in Hu Maxwell, "The Use and Abuse of Forests by the Virginia Indians," *William and Mary Quarterly*, 1st ser., 19 (October 1910), pp. 73-104; Calvin Martin, "Fire and Forest Structure in the Aboriginal Eastern Forest," *Indian Historian*, 6:4 (1973), pp. 38-42, 54; and in Emily W. B. Russell, "Indian-Set Fires in the Forests of the Northeastern United States," *Ecology*, 64 (1983), pp. 78–88. On Indian place-names, see the dictionary by John C. Huden, "Indian Place Names of New England," *Contributions from the Museum of the American Indian*, Heye Foundation, New York, 18 (1962).

The literature on the size of Indian populations at the time Europeans arrived continues to grow. The original estimates were those of James Mooney, "The Aboriginal Population of America North of Mexico," *Smithsonian Miscellaneous Collections*, 80:7 (1928), pp. 1-40. These were very low, but were accepted as authoritative for nearly forty years; they continue to appear in some of the literature. (On Mooney's techniques of estimation, see Douglas H. Ubelaker, "The Sources and Methodology for Mooney's Estimates of North American Populations," in William M. Denevan, ed., *The Native Population of the Americas in 1492*, Madison, 1976, pp. 243-88.) Mooney's figures came under serious attack in Henry F. Dobyns, "Estimating Aboriginal American Populations," *Current Anthropology*, 7 (1966), pp. 395-416, which argued on the basis of disease mortality rates that population estimates should be increased by an order of magnitude or more. Subsequent efforts to revise Mooney drastically upward for the whole continent can be traced in Dobyns, *Native American Historical Demography: A Critical Bibliography*, Bloomington, 1976; Wilbur R. Jacobs, "The Tip of the Iceberg: Pre-Columbian Indian Demography and Some Implications for Revisionism," *William and Mary Quarterly*, 3rd ser., 31 (1974), pp. 123-32; and Douglas H. Ubelaker, "Prehistoric New World Population Size," *American*

Journal of Physical Anthropology, 45 (1976), pp. 661-6. New England estimates can be found in Jennings, *Invasion of America*, pp. 15-31; S. F. Cook, *The Indian Population of New England in the Seventeenth Century*, Berkeley, 1976; and Snow, *Archaeology of New England*, pp. 31-42. On the nature of the epidemics which killed so many Indians, see John Duffy, "Smallpox and the Indians in the American Colonies," *Bulletin of the History of Medicine*, 25 (1951), pp. 324-41; Sherburne F. Cook, "The Significance of Disease in the Extinction of the New England Indians," *Human Biology*, 45 (1973), pp. 485-508; and Billee Hoornbeck, "An Investigation into the Cause or Causes of the Epidemic which Decimated the Indian Populations of New England, 1616-1619," *New Hampshire Archaeologist*, 19 (1976-77), pp. 35-46.

Those wishing to investigate Indian property systems might begin with Imre Sutton's comprehensive bibliography, *Indian Land Tenure*, New York, 1975. As I mention in the endnotes to Chapter 5, the debate about whether Indian individuals and families "owned" hunting territories has been going back and forth since the early twentieth century. The early position was that of Frank G. Speck, who argued (in opposition to a crude Marxist doctrine of primitive communism) that precolonial Algonquian families had in fact owned their hunting territories. See his "The Family Hunting Band as the Basis of Algonkian Social Organization," *American Anthropologist*, 17 (1915), pp. 289-305; "Land Ownership Among Hunting Peoples in Primitive America and the World's Marginal Areas," *Proceedings of the 22nd International Congress of Americanists*, 2 vols., Rome, 1928, pp. 323-32; and, co-authored with Loren C. Eiseley, "Significance of Hunting Territory Systems of the Algonkian in Social Theory," *American Anthropologist*, 41 (1939), pp. 269-80. Important elaborations of Speck's argument can be found in John M. Cooper, "Land Tenure Among the Indians of Eastern and Northern North America," *Pennsylvania Archaeologist*, 8 (1938), pp. 55-9; Cooper, "Is the Algonquian Family Hunting Ground System Pre-Columbian?" *American Anthropologist*, 41 (1939), pp. 66-90; Anthony F. C. Wallace, "Women, Land and Society: Three Aspects of Aboriginal Delaware Life," *Pennsylvania Archaeologist*, 17 (1947), pp. 1-35; and Wallace, "Political Organization and Land Tenure Among the Northeastern Indians, 1600-1830," *Southwestern Journal of Anthropology*, 13 (1957), pp. 301-21. A. Irving Hallowell, "The Size of Algonkian Hunting

Territories," *American Anthropologist,* 51 (1949), pp. 35-45, added an important ecological perspective by arguing that the size of hunting territories depended on the ratio of population to hunting resources. Eleanor B. Leacock, "The Montagnais 'Hunting Territory' and the Fur Trade," *Memoirs of the American Anthropologicial Association,* 78 (1954), called the entire hunting territory concept into question by showing that the system decreased in intensity with increasing distance from centers of the colonial fur trade. Leacock is now ascendant, but see also Dean R. Snow, "Wabanaki Family Hunting Territories," *American Anthropologist,* 70 (1968), pp. 1143-51; Rolf Knight, "A Re-examination of Hunting, Trapping, and Territoriality among the Northeastern Algonkian Indians," in A. Leeds and A. P. Vayda, eds., *Man, Culture and Animals,* AAAS Pub. 78 (1965), pp. 27-42; and R. R. Gadacz, "Montagnais Hunting Dynamics in Historicoecological Perspective," *Anthropologica,* 17 (1975), pp. 149-67. For a more general summary of Indian land concepts drawn principally from later Iroquoian sources, see George S. Snyderman, "Concepts of Land Ownership Among the Iroquois and Their Neighbors," in William Fenton, ed., *Symposium on Local Diversity in Iroquois Culture, BAE Bulletin,* 149 (1951), pp. 15-34. A. L. Kroeber's "Nature of the Land-holding Group," *Ethnohistory,* 2 (1955), pp. 303-14, is based primarily on California sources, but is nevertheless important to this general issue. General histories of the New England fur trade can be found in Francis X. Moloney, *The Fur Trade in New England,* Cambridge, MA, 1931; and William I. Roberts III, *The Fur Trade of New England in the Seventeenth Century,* Ph.D. Thesis, University of Pennsylvania, 1958.

The Colonists

The literature on colonial New England is obviously enormous, but surprisingly little of it speaks directly to ecological issues. Nevertheless, much of ecological importance can be learned from the secondary literature on the colonists' economy, land use, and town organization. The work of historical geographers is particularly useful in trying to reconstruct the spatial organization of colonial settlement patterns. Ralph H. Brown's *Historical Geography of the United States,* New York, 1948, remains the best available general survey of its topic; see also his *Mirror for Americans*, New York, 1943, which describes the East Coast at the beginning of the nineteenth century. H. Roy Merrens surveys the existing colonial literature in his review essay "Historical Geography and Early American History," *William and Mary Quarterly*, 3rd ser., 22 (1965), pp. 529-48; and Douglas R. McManis attempts a new synthesis in his useful *Colonial New England: A Historical Geography*, New York, 1975. Two monographic case studies, neither set in New England, are thorough and suggestive: Andrew Hill Clark, *Acadia: The Geography of Early Nova Scotia to 1760*, Madison, 1968; and Peter O. Wacker, *Land and People: A Cultural Geography of Preindustrial New Jersey*, New Brunswick, NJ, 1975.

General histories of New England that pay some attention to settlement processes include Douglas Edward Leach, *The Northern Colonial Frontier, 1607-1763*, New York, 1966; Charles E. Clark, *The Eastern Frontier: The Settlement of Northern New England, 1610-1763*, New York, 1970; and James A. Henretta's fine synthesis, *The Evolution of American Society, 1700-1815*, Lexington, MA, 1973. Edmund S. Morgan's *American Slavery, American Freedom*, New York, 1975, deals with colonial Virginia, but uses an analysis which holds rich possibilities for comparison with the rest of the colonies. A good general history that focuses especially on different racial and cultural communities in the colonies is Gary B. Nash, *Red, White, and Black*, Englewood Cliffs, NJ, 1974. To place New England in the context of the Atlantic economy generally, three recent texts attempt syntheses: Stuart Bruchey, *The Roots of American Economic Growth, 1607-1861*, New York, 1965; Ralph Davis's elegant review of the major imperial systems of Europe in *The Rise of the Atlantic Economies*, Ithaca, 1973; and the general survey text by Gary M. Walton and James F. Shepherd, *The Economic Rise of Early America*, Cambridge, England, 1979. William B. Weeden's much older *Economic and Social History of New England*, 1890 (reprinted Williamstown, MA, 1978), is a jumble of undigested antiquarian detail, but should nevertheless be examined.

Surveys of New England agriculture are perhaps the best place to begin researching colonial land use. The classic work is that of Percy W. Bidwell and John I. Falconer, *History of Agriculture in the Northern United States, 1620-1860*, Washington, 1925; this should be supplemented by Bidwell's doctoral thesis, "Rural Economy in New England at the Beginning of the Nineteenth Century," *Transactions of the Connecticut Academy of Arts and Science*, 20 (1916), pp. 241-399; and with Bidwell's brief article, "The Agricultural Revolution in New England," *American Historical Review*, 26 (1921), pp. 683-702. Less analytical but still useful for its details is Lyman Carrier, *The Beginnings of Agriculture in America*, New York, 1923; the same is true of Robert R. Walcott's essay, "Husbandry in Colonial New England," *New England Quarterly*, 9 (1936), pp. 218-52. Carl Bridenbaugh, *Fat Mutton and Liberty of Conscience*, Providence, 1974, is especially good on colonial livestock raising. Some of the historical essays in two *USDA Yearbooks of Agriculture* are still worth reading: *Soils and Man*, 1938; and *Climate and Man*, 1941. U. P. Hedrick concentrates on field crops and gardens in his *A History of Horticulture in America to 1860*, New York, 1950. A much more recent synthesis on New England agriculture is as comprehensive as Bidwell and Falconer and should be read, but suffers from antiquarian tendencies and an inadequate analytical apparatus: Howard S. Russell, *A Long, Deep Furrow: Three Centuries of Farming in New England*, Hanover, NH, 1976. To compare colonial agriculture with that of England, Joan Thirsk, ed., *The Agrarian History of England and Wales, vol. IV, 1540-1640*, Cambridge, England, 1967, is essential reading; more popular and more oriented toward material culture is Dorothy Hartley, *Lost Country Life*, New York, 1979. Those interested in how much colonial agriculture exhausted soils should turn to A. P. Usher, "Soil Fertility, Soil Exhaustion, and Their Historical Significance," *Quarterly Journal of Economics*, 37 (1923), pp. 385-411, for a general argument; to Avery Craven's classic *Soil Exhaustion as a Factor in the Agricultural History of Virginia and Maryland, 1606-1860*, in *Illinois University Studies in the Social Sciences*, 13:1 (1926), for a Turnerian analysis of tobacco cultivation in the Old South; and to Edward C. Papenfuse, "Planter Behavior and Economic Opportunity in a Staple Economy," *Agricultural History*, 46 (1972), pp. 297-311, for a refutation of Craven. Warren C. Scoville, "Did Colonial Farmers Waste Our Land?" *Southern Economics Journal*, 20 (1953), pp. 178-81, also attacks the Craven thesis, and uses rather mechanical neoclassical economic assumptions to defend a position that contradicts my own arguments as well.

On the history of colonial lumbering, several essentially antiquarian works contain important information that cannot easily be gotten anywhere else. James E. Defebaugh's *History of the Lumber Industry in America*, 2 vols., Chicago, 1907, is massive and loaded with copious detail. John E. Hobbs, "The Beginnings of Lumbering as an Industry in the New World, and First Efforts at Forest Protection," *Forest Quarterly*, 4 (1906), pp. 14-23, is good on seventeenth-century lumbering practices. Philip T. Coolidge's *History of the Maine Woods*, Bangor, 1963, deals mainly with the nineteenth and twentieth centuries, but its opening sections give useful information on the colonial period. A. F. Hawes, "New England Forests in Retrospect," *Journal of Forestry*, 21 (1923), pp. 209-24, provides a rapid survey of New England forests since the colonial period, and Roland M. Harper, "Changes in the Forest Area of New England in Three Centuries," *Journal of Forestry*, 16 (1918), pp. 442-52, calculates deforestation rates to create a time series for the percentage of cleared land in New England in various periods. Several scholarly studies have focused on English efforts to conserve New England forests for the benefit of the Royal Navy: Robert G. Albion, *Forests and Sea Power: The Timber Problem of the Royal Navy, 1652-1862*, Cambridge, MA, 1926; William R. Carlton, "New England Masts and the King's Navy," *New England Quarterly*, 12 (1939), pp. 4-18; and Joseph L. Malone, *Pine Trees and Politics: The Naval Stores and Forest Policy in Colonial New England, 1691-1775*, Seattle, 1964. Charles F. Carroll, *The Timber Economy of Puritan New England*, Providence, 1973, also considers this issue but embeds it in a larger and more ecologically learned discussion of colonial use of forests generally. Alfred P. Muntz, *The Changing Geography of the New Jersey Woodland*, Ph.D. Thesis, University of Wisconsin, 1964, deals with a non-New England area, but also raises relevant ecological questions. Those interested in investigating what colonists used the forest *for* might begin with Carl Bridenbaugh's rich essay "Yankee Use and Abuse of the Forest in the Building of New England, 1620-1660," *Massachusetts Historical Society Proceedings*, 89 (1977), pp. 3-35. Alex W. Bealer, *The Tools That Built America*, Barre, MA, 1976, is good on colonial carpentry techniques, as are the many books by Eric Sloane. R. V. Reynolds and Albert H. Pierson, "Fuel Wood Used in the United States, 1630-1930," *USDA Circular* 641 (1942), calculate some interesting statistics on firewood consumption, but since their only real data base until the nineteenth century consists of estimates of population size, their numbers should be treated with caution.

To see how agricultural and clearing practices were implemented in local areas, the New England town studies should be examined thoroughly. Many of the older of these pay more attention to land-use practices than do more recent ones. Herbert L. Osgood's essays on land tenure in his *The American Colonies in the Seventeenth Century*, 3 vols., New York, 1904, I, pp. 83-8, 424-67, can still be read with profit. The best early studies are Herbert B. Adams, "Village Communities of Cape Ann and Salem," *Johns Hopkins University Studies in Historical and Political Science*, ser. 1, 9-10 (1883); Melville Egleston, "The Land System of the New England Colonies," *Johns Hopkins University Studies in Historical and Political Science*, ser. 4, 11-12 (1886); Charles M. Andrews, "The River Towns of Connecticut," *Johns Hopkins University Studies in Historical and Political Science*, ser. 7, 7-9 (1889); Nelson P. Mead, "Land System of the Connecticut Towns," *Political Science Quarterly*, 21 (1906), pp. 59-76; and Leonard W. Labaree, *Milford, Connecticut: The Early Development of a Town as Shown in Its Land Records*, New Haven, 1933. Of the later works, Sumner Chilton Powell's *Puritan Village*, Middletown, CT, 1963, with its close examination of how English agricultural institutions were brought to New England and transformed there, remains one of the most suggestive for ecological purposes. It should now be supplemented with David Grayson Allen's more comprehensive *In English Ways*, Chapel Hill, 1981. John R. Stilgoe's *Pattern on the Land*, Ph.D. Thesis, Harvard University, 1976, is excellent on land-use practices, although his notes do not always display the evidence upon which his conclusions rest; his *Common Landscape of America, 1580-1845*, New Haven, 1982, is also very good. Two town studies are particularly insightful on material culture and land use: John Demos, *A Little Commonwealth*, New York, 1970; and Darrett B. Rutman, *Husbandmen of Plymouth*, Boston, 1967. See also Abbott Lowell Cummings, ed., *Rural Household Inventories*, Boston, 1964, which reprints some of the primary documents on which many historians base their analyses of material culture. Other town studies which deal in one way or another with land-use questions include Darrett B. Rutman, *Winthrop's Boston*, Chapel Hill, 1965; Richard R. Bushman, *From Puritan to Yankee*, Cambridge, MA, 1967; Philp J. Greven's excellent *Four Genera-*

tions, Ithaca, 1970; Kenneth A. Lockridge, *A New England Town*, New York, 1970; Lockridge, "Land, Population, and the Evolution of New England Society, 1630-1790," *Past and Present*, 39 (April 1968), pp. 62-80; and James T. Lemon's *The Best Poor Man's Country*, Baltimore, 1972, which deals with southeastern Pennsylvania. Many important essays from this body of scholarship are conveniently reprinted in the two editions of Stanley N. Katz, *Colonial America*, Boston, 1971, 1976. Two architectural studies of town settlement patterns should not be overlooked: Anthony N. B. Garvan, *Architecture and Town Planning in Colonial Connecticut*, New Haven, 1951; and John W. Reps, *The Making of Urban America*, Princeton, 1965.

The colonists' conceptions of property and their justification for dispossessing the Indians are mainly dealt with in older works. A useful general survey is Marshall Harris, *Origin of the Land Tenure System of the United States*, Ames, IA, 1953; Paul Wallace Gates's encyclopedic *History of Public Land Law Development*, Washington, 1968, has only a small section on the colonial period, but is the standard reference on this subject for later periods. Viola F. Barnes, "Land Tenure in English Colonial Charters of the Seventeenth Century," in *Essays in Colonial History Presented to Charles McLean Andrews*, New Haven, 1931, pp. 4-40, is excellent on the medieval origins of the royal charters. Richard B. Morris, *Studies in the History of American Law*, New York, 1930, pp. 69-125, deals at length with colonial land tenure institutions. The origins of the American deed recording system can be explored in George L. Haskins, "The Beginnings of the Recording System in Massachusetts," *Boston University Law Review*, 21 (1941), pp. 281-304; and David T. Konig's excellent "Community Custom and the Common Law: Social Change and the Development of Land Law in Seventeenth-Century Massachusetts," *American Journal of Legal History*, 18 (1974), pp. 137-77. James Sullivan, *The History of Land Titles in Massachusetts*, Boston, 1801, is still useful. Reprinted Indian land titles are conveniently available in Harry A. Wright, ed., *Indian Deeds of Hampden County*, Springfield, MA, 1905; and Sidney Perley, *Indian Land Titles of Essex County, Massachusetts*, Salem, MA, 1912. The litigation which arose between colonists and Indians is lucidly analyzed by James P. Ronda in "Red and White at the Bench: Indians and the Law in Plymouth Colony, 1620-1691," *Essex Institute Historical Collections*, 110 (1974), pp. 200-15.

In addition to the materials discussed in the Indian section above, two articles are good in analyzing the New England colonists' conquest ideologies: Chester E. Eisinger, "The Puritans' Justification for Taking the Land," *Essex Institute Historical Collections*, 84 (1948), pp. 131-43; and Ruth B. Moynihan, "The Patent and the Indians," *American Indian Culture and Research*, 2:1 (1977), pp. 8-18.

The debate over whether colonial farmers engaged in subsistence or commercial agriculture goes back at least to Turner. The works by Bidwell cited above assume that most farmers were self-sufficient and nonmarket-oriented. More recent works which defend similar positions include Cole Harris, "The Simplification of Europe Overseas," *Annals of the Association of American Geographers*, 67 (1977), pp. 469-83; James A. Henretta, "Families and Farms: *Mentalité* in Pre-Industrial America," *William and Mary Quarterly*, 3rd ser., 35 (1978), pp. 3-32; Michael Merrill, "Cash Is Good to Eat: Self-Sufficiency and Exchange in the Rural Economy of the United States," *Radical History Review*, 15 (Winter 1977), pp. 42-71; and Robert E. Mutch, "Colonial America and the Debate about Transition to Capitalism," *Theory and Society*, 9 (1980), pp. 847-63. Authors who argue against colonial self-sufficiency include Rodney C. Loehr, "Self-Sufficiency on the Farm," *Agricultural History*, 26 (1952), pp. 37-41; Darrett B. Rutman, "Governor Winthrop's Garden Crop," *William and Mary Quarterly*, 3rd ser., 20 (1963), pp. 396-415; James T. Lemon, "Early Americans and Their Social Environment," *Journal of Historical Geography*, 6 (1980), pp. 115-31; Winifred Rothenberg, "The Market and Massachusetts Farmers, 1750-1855," *Journal of Economic History*, 6 (1981), pp. 283-314; and Carole Shammas, "How Self-Sufficient Was Early America?" *Journal of Interdisciplinary History*, 13 (1982), pp. 247-72. Bernard Bailyn's classic *The New England Merchants in the Seventeenth Century*, Cambridge, MA, 1955, is also relevant to this controversy. Colonial land speculation as a commercial activity not related to immediate production can be explored in Roy Akagi, *The Town Proprietors of the New England Colonies*, Philadelphia, 1924; Florence M. Woodward, *The Town Proprietors in Vermont*, New York, 1936; Charles S. Grant, "Land Speculation and the Settlement of Kent, 1738-1760," *New England Quarterly*, 27 (1955), pp. 51-71; and Grant, *Democracy in the Connecticut Frontier Town of Kent*, New York, 1961. The relation of property concepts to an emerging ideology of liberal individualism are examined in Richard

Schlatter, *Private Property: The History of an Idea,* 1951; J. E. Crowley, *This Sheba, Self: The Conceptualization of Economic Life in Eighteenth-Century America,* Baltimore, 1974; and Joyce O. Appleby, *Economic Thought and Ideology in Seventeenth-Century England,* Princeton, 1978.

INDEX